National Security and Arms Control in the Age of Biotechnology

National Security and Arms Control in the Age of Biotechnology

The Biological and Toxin Weapons Convention

Daniel M. Gerstein

ROWMAN & LITTLEFIELD PUBLISHERS, INC.
Lanham • Boulder • New York • Toronto • Plymouth, UK

Published by Rowman & Littlefield Publishers, Inc.
A wholly owned subsidiary of The Rowman & Littlefield Publishing Group, Inc.
4501 Forbes Boulevard, Suite 200, Lanham, Maryland 20706
www.rowman.com

10 Thornbury Road, Plymouth PL6 7PP, United Kingdom

British Library Cataloguing in Publication Information Available

Library of Congress Cataloging-in-Publication Data

Gerstein, Daniel M., 1958-
National security and arms control in the age of biotechnology : the Biological and Toxin Weapons Convention / Daniel M. Gerstein.
p. cm.
Includes bibliographical references and index.
ISBN 978-1-4422-2312-7 (cloth : alk. paper) -- ISBN 978-1-4422-2313-4 (electronic)
1. Convention on the Prohibition of the Development, Production, and Stockpiling of Bacteriological (Biological) and Toxin Weapons and on Their Destruction (1972) 2. Biotechnology--Law and legislation 3. Biological arms control. 4. National security--Law and legislation 5. Biological warfare (International law) I. Title.
KZ5825.3.G47 2013
341.7'35--dc23

2013009955

The paper used in this publication meets the minimum requirements of American National Standard for Information Sciences Permanence of Paper for Printed Library Materials, ANSI/NISO Z39.48-1992.

Printed in the United States of America

For those intrepid souls who dare to be more than they thought they could ...

Contents

Acknowledgments   ix

Introduction   xi

1 The Road to the Biological Weapons Convention (BWC)   1
2 Biotechnology and its Implications for the BWC   29
3 Ten Reasons Why the BWC Matters   53
4 Into The Abyss: The Articles of the Convention   85
5 The BWC Review Conferences (RevCon) and Special Reviews   103
6 To Verify or Not to Verify   125
7 The Future of the BWC   145
8 Conclusions and the Way Forward   173

Appendix A: The Biological and Toxin Weapons Convention (BWC)   179

Appendix B: Ten Reasons Why the BWC Matters   185

Appendix C: Confidence-Building Measures   187

Appendix D: Arms Control Definitions   189

Appendix E: Principles of Verification   193

Appendix F: U.S. Ambassador Donald Mahley's Statement   195

Appendix G: National Implementation   207

Notes   211

Index   223

About the Author 233

Acknowledgments

Writing is very difficult for me, and this book almost did not get written. Between work and teaching, time seemed to be the enemy, but then something happened that compelled me to write down these pages.

First, the Biological and Toxin Weapons Convention (shortened to the BWC) review conference, which meets every five years, was held in December 2011 in Geneva. I had a chance to attend and present a briefing on behalf of the U.S. government. Second, every time I decided to leave the topic behind, something seemed to rekindle the fire in me to learn more about the history of biological and toxin weapons and hopefully to contribute to the future of the Convention. And third, friends and co-workers continued to encourage me with articles, comments, and questions about the BWC. The biodefense community, both in the United States and internationally, which is a very small community, some might even say closed, brought me into the circle and continued to fuel my passion. Finally, several of my students at American University came up after one lecture and continued to pepper me with questions about the Convention. It was then I realized that the book must be written.

This book was actually written over a two-year period and reflects my professional journey to learn more about the BWC. When I served in the Countering Weapons of Mass Destruction Directorate of the Office of the Secretary of Defense (Policy), my portfolio included the BWC, and the first overseas trip had me attending a conference at Wilton Park in the United Kingdom dealing with the Convention. It was February 2010, and the review conference was eighteen months away, but the excitement of the international attendees was clear. People spoke of the 2001 debacle, the 2006 review conference that saved the BWC, and the intersessional process that had given new life to the Convention. I even learned that verification was a forbidden word. All of it went right over my head.

Over time and through the encouragement of many friends and co-workers, I began to better understand the BWC, the Articles of the Convention, its shortfalls, and its potential. For this education, I am truly grateful.

In preparing this book, I have had many contributors. For their assistance, suggestions, and patience, I wish to thank Cathy Abbott, Peter Belk, Allison Bender, Tina Carlile, George Christopher, James Dunton, Gigi Kwik Gronvall, Robert Kadlec, David Koplow, Joel McCleary, Jake Meek, Tara Murphy, Tara O'Toole, Ben Petro, Ari Schuler, and Craig Wiener.

Finally, while I am the author of this work, many have made contributions to this effort. These contributions have been immeasurable, and I have worked to ensure that their input is accurately captured. Any inaccuracies in doing so must be attributed solely to me.

Introduction

This book is about biological warfare, or BW, arms control during the Age of Biotechnology.

It examines the continued relevance, even the essentiality, of an arms control treaty—the Biological and Toxin Weapons Convention (BWC)—that has been a foundational element of our nation's security for almost forty years, even eliminating an entire class of strategic and deadly weapons. It is also about an arms control forum that has been meeting just once every five years for decision making on topics that are moving with exponential velocity, and it has reach and implications far beyond the realm of traditional national security and arms control.

In putting together such a book, questions arise, "Who is the target audience?" "To whom will the book appeal?" It is my hope that this book will appeal to a wide audience within the national security, arms control, science and technology fields, and academic disciplines. With the treaty verification debate and the need to reexamine the U.S. position on this topic, those directly involved in developing positions and conducting negotiations should find this book useful. My expectation is that the book will also have broad applicability to people who are interested in security issues but are not professionals in the field. In this regard, the BWC serves as an interesting case study linking the Cold War past with the biotech future. As a result, I believe this book will appeal to those in the biotechnological discipline who are working in biological laboratories attempting to unravel the mysteries of life, to prevent and cure disease, or to prevent a biological warfare (BW) attack from a state actor or bioterrorist. Several examples serve to highlight the types of issues related to the BWC and ultimately to a reader.

In national security terms, we continue to be concerned about a number of nations that have either not joined the BWC or have been deliberately ambiguous about their intentions with regard to biological weapons. In 2013, with the turmoil in Syria, great concern exists concerning the potential stocks of biological weapons Bashar al-Assad is thought to have amassed. Would he use them against his own people or perhaps against an international force? When his regime falls, what will happen to the biological material that he is thought to have? The same questions could be asked with regard to North Korea. Which other nations have stocks of illicit BW?

Even older issues such as the former Union of Soviet Socialist Republics (USSR) program continue to be shrouded in mystery. A new account suggests that President Putin has decried claims that the Soviets ever had such a program, despite overwhelming evidence to the contrary.

Bioterrorism has become an issue of great concern. Three bioterror events serve as harbingers of the potential for a bioterrorist attack: the Rajneeshee cultists in 1984 in Oregon in the United States; attempted attacks by the Aum Shinrikyo in 1990 to 1993 in Japan; and anthrax-filled letters sent to attack a number of targets in the United States in 2001. None of the attacks were particularly effective in causing mortality and morbidity, but their effects transcended their immediate outcomes. In the case of the 2001 attacks that have been labeled Amerithrax, we can measure the result in the billions more in spending directly related to preparedness and response activities that resulted. More recently, al-Qaeda leadership has called for trained microbiologist and chemist jihadists to answer the call to arms.

But the relevance of the BWC is not just to prevent and respond to a state or terrorist BW attack. The reach can and should be much farther. It includes deliberations on issues such as genetically modified organisms (GMO), the response to naturally occurring pandemics, and accidents or biohazards resulting from the misuse of science. In these efforts, the BWC can serve a useful purpose in policy development, coordination with other international organizations and forums, responding to claims of biological events, and assisting nations in developing their readiness against a range of biological threats.

It is ironic that in this Age of Biotechnology defining the early years of this century, some are claiming the Biological and Toxin Weapons Convention (BWC) is becoming irrelevant. The outcome from the 2011 review conference was indeed underwhelming and focused more on the past than on the future. Unfortunately, it saw a return to the discussion of issues that some states had wishfully hoped would remain part of history and that others were determined to resurrect. However, the BWC can and must be much more in this Age of Biotechnology.

In 1969, President Richard Nixon denounced biological weapons and eliminated the U.S. offensive biological weapons program. Much about that decision is known, however, still other aspects of the decision are shrouded in mystery. Why would the president eliminate from the U.S. arsenal weapons that had demonstrated strategic potential? Was he trying to ensure that other nations would follow suit and eliminate what many have come to believe is the "poor man's atomic bomb"? Or was he advised that international negotiations to eliminate biological weapons were on the near horizon and it would make sense to demonstrate U.S. leadership in this area? While the source documentation leading up to the decision has been studied, the final decision calculus remains a mystery. Little has been written about the final decision to renounce offensive biological weapons, and President Nixon made the decision with input from only a small group of close advisors.

Six years later, the BWC was credited with being the first arms control agreement to eliminate an entire class of weapons. In typical arms control fashion, the BWC entered into force with the usual declarations and reservations that accompany arms control treaties. In some regards, the BWC was a standard international agreement with articles of the Convention and allowances for any party to withdraw from the treaty if its supreme interests were jeopardized. Yet one major difference—some might call it a shortfall—separates the BWC from other arms control agreements; the Convention lacked a verification protocol.

This omission was not one of lack of foresight but rather that the parties could not agree to a verification regime or to compliance mechanisms. Perhaps this was the genius of the Convention—allowing for the elimination of a class of dangerous weapons but leaving the details of implementation for a later time. Or perhaps, the framers of the BWC understood that the Age of Biotechnology was on the near horizon and attempts made to "manage" the BWC in the same manner as other arms control agreements would be fruitless. For whatever reason, the BWC has no verification mechanism or measures to provide member-state parties confidence about compliance.

The lack of detail in the original BWC articles—the total Convention is only slightly over 1,700 words—has kept the BWC germane for over thirty-five years. The original articles of the Convention remain as they were at the time of the negotiation of the BWC. Yet in terms of implementation, little has been done to update the Convention. Clarifications have been issued, and modest enhancements, such as the inclusions of "voluntary" confidence-building measures, have been established. Uneven national implementation of the Convention, the lack of universal subscribers to the BWC, and the glacial pace of BWC discussions provides evidence of a lack of progress on even the most basic elements of an arms control treaty.

This comes at a point in the history of humankind that many refer to as the Age of Biotechnology, when advances in medicine, technology, pharmaceuticals, and other related fields are so rapid that laws, policies, regulations, and codes of behavior cannot keep up. Change is occurring exponentially in these key technical fields, and the body of knowledge continues to expand with no end in sight. *De novo* synthesis of living organisms is on the close horizon, genetic manipulations of the very essence of life are occurring with great regularity, and moral and ethic bounds continue to be challenged in the quest for discovery.

While exponential change is occurring in biotechnology, the BWC continues to operate on its own Industrial Age, twentieth-century arms control pace. Periodic reviews held every five years, called review conferences or RevCons, have been augmented with annual meetings of states parties and experts meetings that provide more frequent dialogue on biological issues at the national leadership and technical levels, respectively.

The premise of this book is that the BWC is the most important arms control treaty of the twenty-first century. Its reach goes well beyond what one normally considers for an arms control treaty. From the economy to public health to the environment to national security, the BWC has the potential to impact our lives in very profound ways.

In developing this book, ten reasons form the foundation for this bold assertion about the importance of the BWC. The Biological and Toxin Weapons Convention (BWC) . . .

1. ...eliminates an entire class of dangerous weapons.
2. ...provides an unequivocal norm against the use of biological weapons.
3. ...provides an international forum for dialogue concerning biological defense issues.
4. ...has an important economic dimension.
5. ...is gaining more importance as the spectrum of potential biological threats grows.
6. ...provides a forum for coordinating preparedness and response capabilities against a spectrum of biological threats.
7. ...is an arms control agreement that relates directly to public health, the environment, food security, and biodiversity.
8. ...provides direct linkages to international security mechanisms.
9. ...relates to dual-use capabilities in a way that no other arms control treaty does.
10. ...has responsibilities for implementation that run the gambit from international organizations to the individual.

In this globalized world and having entered into the Age of Biotechnology, we find the BWC at strategic crossroads. Will the Convention continue to muddle through, serving as an observer in the BW debate? Or will it evolve and adapt to meet the challenges posed by the rapid advances in biotechnology?

What is the threat that we face from biological pathogens: state, nonstate, accidental misuse, naturally occurring disease? Does the BWC apply to all of these threats? What is the role of bioethics, biosafety, and biosecurity, and what is their relationship to the BWC?

Is the establishment of the norm against the use of biological weapons through the BWC adequate to ensure that these types of dangerous weapons are never used?

Is a verification protocol or compliance mechanism necessary, as many nations believe, or is it an anachronism of the arms control establishment? What does it even mean to "verify" when change is occurring with such regularity?

Against the backdrop of these questions, this book will consider why the BWC is the most important arms control agreement of the twenty-first century. It will examine the positioning of the convention at the strategic crossroads between national security, arms control, and biotechnology. It is precisely at this intersection that such extraordinary change is occurring with such velocity that policymakers, strategists, biotechnologists, and even ordinary citizens must struggle to understand the implications of the BWC. Major policy changes in biotechnology have the potential to adversely affect U.S. and international biotechnology outcomes. Failure to put appropriate laws, regulations, and policies in place leaves open potential proliferation windows. The need to rely on national and international community cooperation has never been any greater than it is in this space.

This book is also about the history of the BWC, of offensive biological programs of the past and the biodefense efforts of the future. It is also about the nuances of international negotiation and the still-relevant fundamentals of complex arms control treaties.

The first chapter briefly explains the key concepts of arms control and the BWC. It examines the early years of the U.S. offensive biological warfare (BW) program, before the termination of the program by President Nixon in 1969. The chapter also discusses more broadly the history of state BW programs before the BWC period. The important U.S. unilateral decision to renounce is considered in detail, as this both places the rationale for this U.S. unilateral announcement in context and established the United States as a world leader in the anti-BW camp.

The second chapter examines the state of the biotechnological revolution and presents both the path that has been traveled since the BWC entered into force in 1975 as well as the potential for change in this new age. It examines the field of biotechnology, not simply as a single area, but as a collection of technologies that are combined in such a manner as to lead to a multiplicity of outcomes. The early years of the biorevolution have already demonstrated the fundamental changes in parts of society and our daily lives that are possible and have left us with the promise of more to follow. Against this backdrop, the chapter examines the implications for the BWC, given the important transformations that are underway. The transformation that is unfolding before us is the evolution of biology from a largely scientific field to a more industrialized and engineering-based endeavor.

The third chapter develops the argument of the central importance of the BWC to our collective security, public health, economic prosperity, and bio-diversity of our planet. It provides ten reasons why the BWC is important in this emerging Age of Biotechnology. The chapter's purpose is to establish a sense of ownership of the BWC as more than just the realm of a small number of politicians, arms control specialists, and policy wonks, but rather as a treaty that applies to global society and indeed to all individuals.

The fourth chapter examines the BWC in great detail, from the articles of the Convention to the review conferences that are held every five years. It provides an historical analysis of the BWC throughout the almost forty years the convention has been in force. The history of the negotiations, the articles of the Convention, and the review conferences provide critical insights into the BWC. Many of the national positions of today have their roots firmly in the past and suggest strongly the inherent difficulties associated with major changes to the BWC.

The fifth chapter examines the central issue confronting the BWC member state parties now and in the future: the question of verification. The BWC entered into force in 1975 without a verification regime. This omission was understood at the time of signing, yet it did not stop the Convention from gaining 109 state signatories by February 1975. For the next thirty-five years, this question has been revisited and once almost contributed to the demise of the BWC. The chapter concludes by suggesting that opportunities do exist to transform the dialogue on the BWC on the difficult verification issue; the question will be whether the parties to the convention can find the will to do so.

The sixth chapter provides a prescriptive analysis of the BWC looking for ways to improve aspects of the treaty, including through the development of a verification protocol. This must be done with great care and incrementally so as not to do something that would destroy the convention or have a deleterious effect on the biodefense community, a very delicate balance.

The final chapter provides conclusions and thoughts for the future. It is not intended to be a cookbook with recipes for broad, sweeping changes to the Convention. That would be presumptuous and fraught with a very high likelihood of widely missing the mark. Rather, the chapter provides a struc-tured way of thinking about the future of the BWC.

The book includes reference material that I hope will be useful to the reader, including the text of the BWC; the confidence-building measures form categories; definitions of arms control and verification; principles of verification from the United Nations; the text of the U.S. ambassador's speech to the 2001 review conference in which the United States walked out of the negotiations; and some statistics on national implementation.

Negotiations by their very nature are difficult, complex, and require the willingness to find solutions where none appear to exist. They are also not for the faint of heart and require patience and compromise. Negotiations can also not be approached as an all-or-nothing proposition or as requiring complete surrender. The art of the deal is finding a solution that is good enough for all parties, or at least one that all parties can live with. This spirit is embodied in Paul Nitze's (a former Deputy Secretary of Defense) concept of "effective verification."

It is also not my intention to radically change the U.S. position regarding the BWC, in particular the verification issue. Rather, my thought would be to ensure that the Convention does not drift into irrelevance by failing to address the issues that are perceived by many BWC member states to be central to the future of the Convention. Alternatively, we must ensure that any changes to the Convention do not have second- and third-order deleterious effects. To this end, we must embrace the BWC as a different type of arms control treaty and avoid the temptation for identifying easy answers; if there were easy answers remaining, they would have already been identified and addressed.

Chapter One

The Road to the Biological Weapons Convention (BWC)

Arms control is an arcane subject. It is practiced by a relatively small subsection of the security and defense community. Many times, these individuals are heard being referred to as arms control wonks, interchangeably in both a reverent and disparaging manner. Arms control has important implications for the lives of daily citizens, but it has little visibility with the average person. People do not normally give a second thought to controlling activities at sea, limiting nuclear weapons, or banning land mines. But these issues, along with other arms control treaties, are important to the national and homeland security of the United States, and the global community and its citizens.

And if arms control is arcane, then the Biological and Toxin Weapons Convention (BWC) is surreal. The BWC is a convention that has implications for a sector of the U.S. economy that is approaching 20 percent of the Gross National Product (GNP) and includes the collective fields of biotechnology, medicine, public health, and agriculture. It is also an agreement that few outside of a relatively small community know exists. Even many scientists who work within these fields are largely unaware of the BWC or the collective requirements that it portends.

At its very essence, arms control is about negotiation. First, there is the negotiation about what implements of war are to be included in the treaty or what behavior is to be controlled. Second, an agreement must be reached concerning the basic provisions of the treaty and how it will be implemented and monitored. Third, the "finalized" treaty must be sent back to capitals where it must be agreed to so that the treaty can enter into force (EIF). Finally, there is the negotiation that takes place on a periodic basis between policymakers seeking to "monitor" the implementation of the treaty or on the ground as inspectors seek to gain an understanding of the compliance of the treaty.

Arms control has come to mean different things to different people, and its meaning has evolved throughout history. Arms control agreements existed prior to World War II (1939–1945); however, the modern arms control effort began in earnest after the Cuban Missile Crisis of 1962.[1] During the postwar period, the theory and practice of modern arms control came into being. The focus was on dealing with the mistrust between the United States and Soviet Union. Arms control was seen as a way to limit holdings of weapons systems, to control behavior, and to increase transparency and stability.

Today, our concept of arms control has evolved again to include such activities as nonproliferation and the Cooperative Threat Reduction (CTR) program. And instead of a focus on managing U.S.-Soviet (or -Russian) relations, arms control has become universal with treaties intending to involve the global security community. While the BWC was developed during the height of the Cold War, one could argue that it was ahead of its time in being one of the first treaties to include all nations desiring to participate, and it was certainly the first of the Chemical, Biological, Radiological and Nuclear (CBRN) treaties to have, by necessity, broad international membership.

Of course, arms control is a two-edged sword and as such, it cuts both ways. Transparency and openness into the affairs of another nation can translate into loss of security and even infringement on a state's sovereignty. It means that while one nation might be able to gain insights into the activities of another nation, an expectation of reciprocity accrues.

The philosophy of arms control is derived from the belief that a nation's security is enhanced through entering into an arms control agreement. In such a case, "a country or countries restrict the development, production, stockpiling, proliferation, distribution or usage of weapons. Arms control may refer to small arms, conventional weapons or weapons of mass destruction (WMD) and is usually associated with bilateral or multilateral treaties and agreements."[2]

For this security assurance, nations agree to trading portions of their national sovereignty for the enhanced security that accompanies the treaty through the limitation of a behavior or the reduction or elimination of a perceived threatening capability. Compliance with the arms control agreement is normally verified by on-site inspections, satellites, and/or overflights by airplanes. Additionally, arms control agreements normally include provisions for data exchanges, challenge inspections, and the reporting of exercises as transparency measures. Collectively and by way of shorthand, these are normally referred to as verification.

The agreement and implementation creates another important characteristic of arms control agreements: the ability to withdraw from a treaty based on supreme national interest. This provision normally requires both notification of the member party's intent and a time frame for withdrawal. When entering into an arms control treaty, nations have come to understand that conditions may arise or changes might occur that tip the balance away from the belief that the agreement is resulting in mutually enhanced security and toward concerns that security is being diminished through continued participation in the treaty.

Nuclear arms control treaties undoubtedly get the most press and have the greatest following. An entire community of strategic experts has grown up talking about the strategic balance of power, missile delivery systems, and throw weights. The alphabet soup of nuclear treaties, or those dealing with nuclear-related issues, is long and complex, with equally long and complex documents for implementation and verification.

The Strategic Arms Limitation Talks (SALT I and SALT II) agreements between the United States and the Soviet Union were the first to place limits and restraints on some of the central and most important armaments, strategic offensive weapons. These agreements resulted in the limitations to their most powerful land- and submarine-based offensive nuclear weapons. In what was considered a major breakthrough at the Vladivostok meeting in November 1974 between President Ford and General Secretary Brezhnev, both sides agreed to a basic framework for the SALT II agreement that included a 2,400 equal aggregate limit on strategic nuclear delivery vehicles (Intercontinental Ballistic Missiles (ICBM), Submarine Launched Ballistic Missiles (SLBM), and heavy bombers), as well as limiting multiple integrated reentry vehicles (MIRV) systems; a ban on construction of new land-based ICBM launchers; and limits on deployment of new types of strategic offensive arms. Verification would have been accomplished by "national technical means (NTM) of verification, including photo-reconnaissance satellites." Furthermore, each agreed not to interfere with each other's national technical means, and not to use deliberate concealment measures that would have impeded verification by NTM of compliance with the provisions of the agreement. SALT II never entered into force based on political considerations at the time.

Later, important nuclear treaties sought to reduce and eliminate nuclear capabilities and to increase transparency and confidence between the signatories. The Intermediate-Range Nuclear Forces (INF) Treaty between the United States of America and the Union of Soviet Socialist Republics eliminated intermediate-range and shorter-range missiles. The president ratified this treaty on May 27, 1988, after the advice and consent of the U.S. Senate.

The Strategic Arms Reduction Treaty (START with I and II) reduced and limited intercontinental ballistic missiles (ICBMs) and ICBM launchers, submarine-launched ballistic missiles (SLBMs) and SLBM launchers, heavy bombers, ICBM warheads, SLBM warheads, and heavy bomber armaments. The Strategic Offensive Reductions Treaty (SORT) continued with further reductions in strategic offensive weapons in 2002. Most recently, New

START signed in April 2010 is a U.S.-Russian treaty that should result in further reductions in nuclear weapons holdings.

There are also treaties on Nuclear-Weapon-Free Zones (NWFZ), which are regionally based and are designed to limit nuclear activities within the areas of application. They have strange names such as the Treaty of Pelindaba for Africa or Treaty of Tlatelolco for Latin America. In all there are five NWFZ treaties, some of which the United States is party to and others that the United States has not agreed to participate in.

Other treaties related to limiting nuclear capabilities include the Nuclear Non-Proliferation Treaty (NPT), which "undertakes not to transfer to any recipient whatsoever nuclear weapons or other nuclear explosive devices or control over such weapons or explosive devices directly, or indirectly; and not in any way to assist, encourage, or induce any non-nuclear weapon State to manufacture or otherwise acquire nuclear weapons or other nuclear explosive devices, or control over such weapons or explosive devices," as well as providing safeguards through inspections by the International Atomic Energy Agency (IAEA). Some treaties such as the Comprehensive Nuclear-Test-Ban Treaty (CTBT) have yet to be ratified by the president, although they have essentially (by practice) been provisionally applied.

Still, there are other arms control treaties and regimes that limit non-nuclear activities and weapons. The Conventional Armed Forces in Europe (CFE) Treaty limited conventional weapons holdings of tanks, artillery, infantry fighting vehicles, helicopters, and fixed-winged aircraft in a region of the former Soviet Union, former Warsaw Pact nations, and the sixteen members of the North Atlantic Treaty Organization (NATO) in 1991. The Missile Technology Control Regime (MTCR) limits the proliferation of missile technology to systems not capable of delivering "at least a 500 kg [kilogram] payload to a range of at least 300 km [kilometers] as well as the specially designed 'production facilities' for these systems." The Outer Space Treaty governs the activities of states in the exploration and use of outer space, including the moon and other celestial bodies. The Mine Ban Treaty and Convention on Cluster Munitions are two like-purposed treaties being considered for responding to a growing humanitarian crisis caused by unexploded munitions around the globe; both have entered into force.

Perhaps the treaty most closely related to the BWC—at least by history and in the minds of many experts—is the Chemical Weapons Convention (CWC). Both derive from the Geneva Protocol of June 17, 1925, known as the Protocol for the Prohibition of the Use in War of Asphyxiating, Poisonous or Other Gases and of Bacteriological Methods of Warfare, which entered into force on February 8, 1928.

The CWC eliminates all chemical weapons as weapons of war. To ensure continued application of the convention, the language has been made broad and based on the intent or purpose of the state rather than on specific lists of prohibited chemicals and precursors. The treaty includes three schedules or lists of toxic chemicals and precursors. To demonstrate the resilience of the CWC language, one need only consider that the schedules of chemicals to be considered under the CWC have not been changed since the treaty entered into force in 1997.[3] However, the CWC contains several interesting exclusions, including the use of toxic chemicals for industrial, agricultural, research, medical, pharmaceutical, and other peaceful applications; the development of defensive countermeasures; and law enforcement, including domestic riot control. This exclusion has led to significant ambiguity with respect to the ongoing activities of several of the member states. In contrast to the BWC, the CWC does contain provisions for verification and treaty compliance.

Article I, General Obligations, of the Chemical Weapons Convention requires,[4]

1. Each State Party to this Convention undertakes never under any circumstances:

   a. to develop, produce, otherwise acquire, stockpile or retain chemical weapons, or transfer, directly or indirectly, chemical weapons to anyone;
   b. to use chemical weapons;
   c. to engage in any military preparations to use chemical weapons;
   d. to assist, encourage or induce, in any way, anyone to engage in any activity prohibited to a State Party under this Convention.

2. Each State Party undertakes to destroy chemical weapons it owns or possesses, or that are located in any place under its jurisdiction or control, in accordance with the provisions of this Convention.
3. Each State Party undertakes to destroy all chemical weapons it abandoned on the territory of another State Party, in accordance with the provisions of this Convention.
4. Each State Party undertakes to destroy any chemical weapons production facilities it owns or possesses, or that are located in any place under its jurisdiction or control, in accordance with the provisions of this Convention.
5. Each State Party undertakes not to use riot control agents as a method of warfare.

The CWC is supported by a body that performs the policy, verification, and administrative requirements of the Convention. The Organization for the Prohibition of Chemical Weapons (OPCW) is a standing body that meets regularly in The Hague, Netherlands, to discuss topics of interest and significance to the implementation and continued relevance of the Convention. Of particular significance is the continued work of the OPCW in the oversight of the destruction of the chemical stockpiles from the Cold War in Russia and the United States; an on-site OPCW team verifies all destruction of munitions, and regular reports are made on destruction progress. The CWC has major review conferences that are held approximately every five years.

So that brings us to the Biological and Toxin Weapons Convention, which as described before, is surreal. Even the name is confusing. Some call the treaty the BTWC, while others shorthand it to the BWC, leaving out the reference to toxins. Provisions of the Convention call for a meeting of states parties (MSP) on an annual basis, with a formal review conference or RevCon every five years. No decisions are undertaken at the MSP; those are reserved for the RevCon. The RevCon is a three-week meeting in Geneva that seemingly accomplishes very little. The first week includes national statements of resolve and hope for reinvigoration of the Convention. During the second week, positions begin to become staked out and hardened; alliances are formed. The third week is dedicated to arriving at a consensus document, which, as history has demonstrated, advances the Convention very little, if any.

The language of the BWC is equally instructive. It is a short, legally binding, politically based statement of intent rather than a "normal" arms control treaty with tens or even hundreds of pages of supporting documentation that delineates the limits of the agreement. The Convention, which has not been changed from the original language, is comprised of fifteen articles that fit on five or so pages. It provides no definition of and no mechanism for assessing compliance. The centerpiece of the Convention is Article I, which states:[5]

> Each State Party to this Convention undertakes never in any circumstances to develop, produce, stockpile or otherwise acquire or retain:
>
> 1. Microbial or other biological agents, or toxins whatever their origin or method of production, of types and in quantities that have no justification for prophylactic, protective or other peaceful purposes;
> 2. Weapons, equipment or means of delivery designed to use such agents or toxins for hostile purposes or in armed conflict.

Comparing the language in the CWC and BWC provides an interesting contrast and provides a basis for understanding the similarities in the two conventions. Both rely on broad purpose or intent-based language that is less about absolute lists than about the intended use of the material. Both rely on national implementation to ensure the provisions of the respective conventions are adhered to. Both also rely on periodic review conferences to assess the continued relevance of the conventions. Of course, the major glaring difference between the two conventions is the lack of a verification protocol in the BWC.

Some important overlap exists between the CWC and the BWC. Toxins produced by living organisms are covered by both the CWC and BWC. For example, two toxins—saxitoxin and ricin—are listed on Schedule 1 of the CWC and are therefore subject to verification inspections. Later, we will discuss how technological advancements are leading to a natural convergence of the CWC and BWC given the advent of synthetic biology.

The BWC—as we refer to it in the United States—was signed in Washington, London, and Moscow on April 10, 1972; gained advice and consent by the U.S. Senate on December 16, 1974; was ratified by the U.S. president on January 22, 1975, with ratification deposited at Washington, London, and Moscow on March 26, 1975; was proclaimed by the U.S. president on March 26, 1975; and entered into force on March 26, 1975.

BIOLOGICAL WARFARE

Biological warfare (BW) is the use of biological material or derivatives of biological material as a weapon. In short, it is the use of microbes to deliberately cause disease. But herein lies the complication. Microbes are inherent in all life forms and include bacteria, viruses, protozoa, algae, and fungi. These forms of biological material have great importance in all matter of life on Earth. In the appropriate mix, microbes play a vital role in normal living processes, including digestion and the environment. One account describes the abundance of microbes as follows:

> Microbes have dominated Earth for more than three billion years and are the basis for all other life-forms. Microbes account for more than 60 percent of all the Earth's organic matter and weigh more than 50 quadrillion metric tons. Microbes are everywhere and constitute nine out of ten cells in the human body. There are uncounted millions of different kinds of microbes, but only a few thousand cause disease in plants and animals.[6]

In biological warfare, we are only concerned with an even smaller number of pathogens that are both known to cause disease and can be weaponized so that they cause disease efficiently and with predictable results.

Developing biological weapons includes many significant factors. First is the choice of agents; it is generally a bacteria, virus, or toxin, although fungi and rickettsia have been contemplated for use as biological weapons. In considering the choice of agent, one must also consider the question of whether a contagious or noncontagious pathogen is desired and whether an incapacitating or lethal weapon should be incorporated. The selection of a pathogen must be linked to the desired outcome that is intended. On the other hand, at an overwhelming dose or for an immunocompromised person, even an incapacitating agent can be lethal. The range of pathogens is wide, and therefore the number of organisms or biological material required to have a lethal effect for half the exposed population, normally expressed as lethal dose 50 or LD50, can vary greatly. For example, the LD50 for anthrax is eight thousand organisms, for tularemia it is fifty bacteria, and for the Ebola virus is ten virions. While much of the focus of countering BW efforts falls on the pathogens, producing a BW capability also relies on several key factors.

The pathogen must be appropriately grown and manufactured through the culturing, milling, and drying process to ensure that it is highly virulent. The formulation of the pathogen, to include whether it has been stabilized for release in the environment and had additives that can affect both viability and decay rates, are key factors as well. The deployment method—aerosol, vector-borne, food- or water-borne, or through explosive munitions—must be considered within the context of the desired effect. A biological attack is also highly affected by the meteorological and terrain within which the weapon is released. All of these factors combine to provide an indication of the biological weapon efficiency.

The effectiveness of a biological warfare attack is also related to the concentration of the pathogenic material. If the concentration is low, the likelihood of a successful attack is also low, while if a high concentration of the pathogen is achieved, the chances of a successful attack greatly increases. The dose of the pathogen that each individual receives also plays an important role in the effectiveness of an attack. Dose can be affected by such individual factors as respiration rate and even position in relation to the aerosol release. Two people standing next to each other can receive vastly different dosages. Target susceptibility also plays a vital role in an attack. Is the population naïve? Have they been vaccinated? Are medical countermeasures available?

The complexity inherent in the BWC lies in being able to take the very general provisions of Article I of the Convention and apply them to this very complex marriage of biological material, biotechnological capabilities, and knowledge to ensure that the unequivocal ban on BW is being adhered to in the letter and spirit of the Convention.

The history of BW in conflict goes back to the antiquities. In fact, BW was used as a means of attacking one's enemy and inflicting death long before the concept of disease was understood. In these early times, only the cause and effect was understood, not the mechanisms of action whereby disease was spread. What was understood was that if an arrow dipped in feces hit an enemy soldier, a sickness would be inflicted upon the victim that would ensure that the wound, no matter how severe, was likely to result in death. In the same way, blankets used by smallpox victims could be provided to noninfected people and cause the deadly disease to spread. However, the understanding of disease would come much later.

Scientific discovery beginning in the 1800s led to the body of knowledge concerning the spread of disease. Scientific discovery in Europe by such giants as Edward Jenner, Louis Pasteur, Robert Koch, and John Snow would begin to unravel the mysteries of disease. Jenner's observations and theories led to the first vaccinations of healthy people with the *vaccinia* or cowpox virus to inoculate against the dreaded smallpox disease; his work was widely regarded as the foundation of immunology.[7] Pasteur would develop a process to cleanse milk of the dangerous microbes in a process that became known as pasteurization. Koch developed what became then and remains today the standard for understanding disease, Koch's Principles. And Snow came to understand that cholera was not caused by vapors but rather by a microscopic organism associated with water contaminated with sewage that was laced with the *Vibrio cholerae* bacterium. These were serious scientists seeking to understand disease at a very different level than the early bioweaponeers merely attempting to defeat their adversaries using any and all available means.

In the United States, the understanding of disease progressed more slowly. It was not until the 1918 Great Influenza pandemic that medicine began the progression from anecdotal treatment of disease to gaining an understanding of disease at the molecular level.

THE QUESTION OF THE U.S. PROGRAM

The decision to renounce BW remains somewhat mysterious. It was made by a small group of national decision makers who reported directly to the president. It came at a time when the U.S. offensive program had concluded a series of successful tests that demonstrated the utility of BW.

The debate leading up to the 1969 Nixon decision had been ongoing since the 1925 Protocol for the Prohibition of the Use in War of Asphyxiating, Poisonous or Other Gases, and of Bacteriological Methods of Warfare.[8] At various times the discussion gained more or less visibility depending on complementary debates about the use of force and protection of populations against perceived threats. During World War II as part of such a policy debate, President Roosevelt denounced the use of such weapons. However, as part of the discussion, Roosevelt made clear that the restraints did not pertain to defensive preparations. Roosevelt had commissioned a study in 1941 that was made public in 1946 concerning biological weapons. The Merck Report, named after its chairman, George W. Merck, was charged with investigating the potential of biological warfare. One account summarizes the results as reminding people that nuclear weapons were no longer the only weapons capable of dealing death and disability in catastrophic proportions.[9]

In 1947, the Federation of American Scientists (FAS) called for the international control of biological weapons. The plea was supported by a former member of the Airborne Infection Laboratory at Fort Detrick, Dr. Theodor Rosebury, who had worked on biological weapons during World War II. With the allegations of U.S. use of biological weapons in the Korean War, the debate took on additional importance. One summation from the FAS stated, "Anyone who would deliberately further the spread of disease is in league with evil, feared, and little understood forces."[10]

Within the Armed Forces, supporters of the development of biological (and chemical) weapons argued for and even tried to convince the public and decision makers that biological weapons were an acceptable form of warfare. These efforts were blunted with opposing dialogue, including legislation presented by Congressman Robert W. Kastenmeier, calling for a "no first use" policy concerning biological weapons. Both the departments of State and Defense opposed the legislation, both on practical as well as on philosophical terms. Practically speaking, the government had come to understand the potential of biological weapons, while in philosophic terms, no first-use declaratory policies were not favored.

International debate continued with one particularly noteworthy discussion at the 1959 Pugwash Conference (also known as the Conference on Science and World Affairs). Members from NATO, the Warsaw Pact, and the nonaligned nations met to consider the potential for chemical and biological weapons control. Of interest was the difficulty noted in "maintaining control over these agents in the field," a harbinger of the verification debate associated with the BWC at the time of its crafting and today.[11]

In the United States, the question remained unsettled, and preparations for offensive and defensive biological warfare continued. At times, conflicting messages were sent at the highest levels. During the Eisenhower Administration, Basic National Security Strategies, or BNSPs, provided strategic guidance on key issues. The first reference to biological weapons was in a 1956 BNSP, which stated, "To the extent that the military effectiveness of the armed forces will be enhanced by their use, the United States will be prepared to use chemical and bacteriological weapons in general war. The decision as to their use will be made by the President."[12] During a February 1960 meeting of the National Security Council, President Eisenhower was briefed again about potential scenarios for the use of incapacitating chemical and biological agents, and he was quoted in the meeting notes as stating it was "a splendid idea," yet he cautioned that there was potential for retaliation in kind.[13] Later in that same year when President Eisenhower was asked about biological weapons in a press conference, he responded, "I will say this: no such official suggestion has been made to me and so far as my own judgment is concerned, is not to start such a thing as that, the use of chemical or biological weapons first."[14]

During the Kennedy Administration, with the development of the concept of "flexible response," the potential for controlling chemical and biological weapons had been further lessened. Combined with the increasing use of defoliants during operations in Southeast Asia, thresholds against the use of chemical and biological capabilities had been further lowered.[15] Use of these agents (note that they were all chemical) was not without some controversy and discussion. One item of particular importance was the use of defoliants to destroy crops in a sort of biological warfare using chemical agents. These debates were heightened during the end of the Johnson administration during the period of 1965 to 1968. One anecdote that highlights the significance of the concern was the presentation by the president's science advisor in 1967 of a letter signed by over five thousand scientists calling for the halting of the use of chemical weapons in Vietnam.[16]

While these increasingly acrimonious debates were primarily dealing with the chemical weapons issue, the relationship between chemical and biological weapons had been established authoritatively in the 1925 Protocol and continued within policy debates and even through organizational relationships such as the assignment of responsibility for both offensive and defensive chemical and biological issues being in the U.S. Army Chemical Corps which further contributed to this alignment.

The argument on biological weapons was sharpened by a 1966 statement from Nobel Laureate geneticist Joshua Lederberg from Stanford University, who was pressing for an international agreement on biological weapons. His rationale was as follows:

> Its development is closest to medical research, therefore conveys the most intense perversions of the human aims of science. It is the most dubious of military weapons. Its effects in field use are most unpredictable, with respect to civilian casualties, and even retroactive on the user. The large scale development of infectious agents is a potential threat against the whole species: mutant forms of viruses could well develop that would spread over the earth's population for a new Black Death.[17]

While Lederberg's arguments were compelling, they were not universally accepted, and the debate continued. The U.S. military consistently fought to retain the flexibility to develop both offensive and defensive BW capabilities. A more complete discussion of these efforts will be provided later, yet suffice it to say that the strategic potential for biological weapons was becoming well understood. However, there was also the realization among military planners and civilian leadership that the possession of nuclear weapons that were designed to kill populations made biological weapons that had the same motivation a redundant capability.

However, within the international community, the tide was beginning to turn against chemical and biological weapons. At the 1966 meeting of the

United Nations General Assembly (UNGA), a draft resolution on the use of chemical and biological weapons was presented. The resolution was primarily related to U.S. actions in Vietnam regarding the use of defoliant. The U.S. ambassador to the UNGA made the statement that the United States would not be the first to use lethal chemical and biological weapons. The meaning of this statement was ambiguous even to U.S. policymakers, who were unclear as to whether this meant a renunciation of the use of only lethal agents or all biological agents.[18]

In the spring of 1968, the Eighteen Nation Disarmament Committee (ENDC) met to conclude negotiations on the Nuclear Nonproliferation Treaty. At the behest of the U.K. delegation, the inclusion of chemical and biological weapons control was raised. And it was during this requested inclusion that the bifurcation of the chemical and biological weapons began to occur. Since the 1925 Protocol for the Prohibition of the Use in War of Asphyxiating, Poisonous or Other Gases, and of Bacteriological Methods of Warfare, chemical and biological weapons had been considered together; however, international discussions focusing on U.S. actions in Vietnam began to unravel this historical linkage.

As a result, biological and chemical weapons would be limited by different conventions—the BWC and CWC—and separated in time by entry into force of 1975 and 1993, respectively. It is within this backdrop that the Nixon decision on biological weapons was made.

THE UNITED STATES' REVIEW OF ITS OFFENSIVE BW PROGRAM—NATIONAL SECURITY STUDY MEMORANDUM (NSSM) 59

In November 1969, U.S. President Richard Nixon renounced the use of biological weapons. This announcement followed a multiagency—today we would call it an interagency—review of chemical-biological warfare (CBW) policies as part of National Security Study Memorandum (NSSM) 59. The NSSM required the secretaries of State and Defense, the director of the Arms Control and Disarmament Agency (ACDA), the director of the Central Intelligence Agency (CIA), and the Chairman of the Joint Chiefs of Staff (CJCS) to examine the CBW policies in a methodical manner to determine the utility of these efforts.[19]

The seeds of this review began several months after President Nixon's inauguration with the short, terse letter in April 1969 from the Secretary of Defense Melvin Laird to the Assistant to the President for National Security Affairs, Dr. Henry Kissinger, requesting a policy review of the chemical and biological programs, as it was "clear the Administration [was] going to be under increasing fire as a result of the numerous inquiries, the more notable being Congressman McCarthy's and Senator Fulbright's."[20]

Additionally, at this same time, several events became conflated that served to act as an accelerant for calls for a review of the U.S. chemical and biological programs. The linkage of the two programs in many policy documents and organizationally by the U.S. Armed Forces after six thousand sheep were exposed to a powerful nerve agent at Dugway Proving Ground in Utah (with the resulting death of over three thousand sheep and the contamination of a large area of the surrounding community) and the coincidental discovery of the U.S. Army's plan to dump chemical munitions into the ocean led to calls for congressional inquiries and administration deliberations on the issue. A key figure in the calls for disarmament was Dr. Matthew Meselson, who testified in a closed hearing on the general subject of chemical and biological disarmament and through his relationship with Dr. Kissinger had provided position papers on the topics to the administration. It was under that backdrop that the Nixon Administration undertook a comprehensive review of the U.S. chemical and biological programs.

NSSM 59, dated May 28, 1969, required the participants to examine current programs, to identify issues, and to develop alternatives. Specifically,

> the analysis should delineate (1) the nature of the threat to the U.S. and its Allies and possible alternatives in meeting this threat; (2) the utility of and circumstances for possible employment of chemical and biological agents, both lethal and incapacitating; (3) the operational concepts relating to possible use, testing and stockpiling; (4) the research and development objectives; (5) the nature of and alternative approaches to the distinction between lethal and non-lethal chemical and biological agents, including a review of current applications of U.S. policy relating to non-lethal agents such as chemical riot control agents and chemical defoliants; and (6) the U.S. position on arms control, including the question of the ratification of the Geneva Protocol of 1925.[21]

NSSM 59 required the staffing of three interagency groups. The first was composed of the Intelligence Community (IC), with the role to assess foreign chemical and biological capabilities. The second, composed primarily from representatives of the Department of Defense (including the JCS), had the task of examining the military options available to the president and in particular, the military utility of chemical and biological weapons. The third, composed from across the interagency, had the task of examining diplomatic options considering the ongoing international initiatives. Given the composition of the three groups, it comes as little surprise that significant differences arose nearly immediately. It became evident that even within a single department, little consensus could be reached. In the Department of Defense, for example, one briefing paper prepared by the Office of Systems Analysis criticized biological weapons as to their battlefield capability and lack of deterrent value, while a second report prepared by the JCS highlighted their "reliability and controllability." In response to the JCS report, Defense Secretary Laird had the JCS members removed from the Defense Interagency group.[22]

Simultaneously, the NSC requested the Office of Science and Technology prepare a report on chemical and biological weapons technology that was to serve as a primer for the NSC staff and an introductory document on the issue for Dr. Kissinger, the president's National Security Advisor. The report was to be written by members of the President's Science Advisory Committee (PSAC). This committee concluded that the United States: (1) should give up its offensive BW capabilities, mothball existing facilities, and continue to develop defensive capabilities; (2) ratify the 1925 Geneva Protocol as a means to improve the U.S. image internationally; (3) renounce first use of lethal and incapacitating agents; (4) undertake additional studies concerning the long-term implications of using herbicides and riot control agents; and (5) continue research on the synthesis of toxins.[23]

The NSSM 59 memorandum required participants to consider the effects on U.S. "international posture" and "upon relationships with Allies." The entire review effort and group conclusions were to be forwarded to the National Security Council (NSC) by September 5, 1969, a little over three months. Thus began a series of weekly, high-level deliberations to consider the fate of the U.S. offensive biological and chemical programs.

The international handwriting was on the wall, clearly signaling the likely direction of biological warfare. Simultaneous to the discussions that were ongoing as a result of NSSM 59, the United Kingdom had been contemplating the idea of banning biological weapons. In July 1969, the United Kingdom formally submitted a proposal to the Committee on Disarmament (CD) of the United Nations prohibiting the development, production, and stockpiling of biological weapons, and providing for inspections in response to alleged violations. In August 1969, Sweden submitted a draft condemning and declaring contrary to international law any use of biological (and chemical) weapons in armed conflicts. Canada recommended that all states accede to the 1925 Protocol and used the United Nations Secretary General (UNSYG) report as a rationale for the elimination of biological weapons.[24] In September 1969, Soviet Foreign Minister Andrei Gromyko, in a speech to the United Nations General Assembly, proposed a treaty banning the development, production, and use of both chemical and biological weapons; the linking of the chemical and biological weapons was intended to use the momentum gathering in the BW arena to address the CW issue, where the Soviets believed a U.S.-Soviet arms race was underway.[25]

NSSM 59 dealt with both chemical and biological weapons, but for the purposes of this work, only the BW discussions and implications will be considered. The study began by examining the characteristics of biological warfare that distinguished it from other types of warfare. For example, biological weapons required an incubation period that varied by pathogen for the effect of these weapons to be seen, and therefore, it perhaps limited their utility as a battlefield weapon. The study also noted the inability to definitively target, thus leading to the potential for inadvertent exposures to civilian populations or other inadvertent targets.

The NSSM considered the potency of biological weapons, noting the relatively small amount of material required to have a large effect. On the issue of the use of biological weapons, the formerly top secret NSSM concluded, "As far as is known, biological agents of warfare have never been employed in modern times."[26] This last conclusion is interesting, as the statement does not account for the employment by the Germans in World War I of biological agents to infect and kill horses and other draft animals or the use of biological agents and experimentation during World War II by Japanese Unit 731 against the Chinese population and enemy prisoners of war.

In discussing the international state-of-play and the position of the USSR in particular, the NSSM final document reports that before a United Nations General Assembly (UNGA) on September 19, 1969, "Foreign Minister Gromyko announced that the USSR, all other Warsaw Pact countries (except East Germany), and Mongolia were submitting to the 24th UNGA an item on the conclusion of a convention on the prohibition of the Development, Production and Stockpiling of Chemical and Biological Weapons and on their destruction."[27] These provisions were intended to apply to both chemical and biological weapons. In a harbinger of things to come, the USSR proposal relied on "self-policing rather than on any plan of international control by inspection."[28] Interestingly, the USSR proposal largely survived within the BWC final text and formed the basis for Article I of the Convention. Additionally, the proposal for self-policing rather than the development of an intrusive international verification regime became an important part of the history and criticism of the Convention that continues to be part of the BWC's landscape today.[29]

In a memorandum dated October 22, 1969, addressed to National Security Advisor Henry Kissinger, the President's Science Advisor, Lee Du-Bridge, discussed the three major recommendations and his support for them. NSSM 59 had recommended the renouncing of all offensive BW, stopping completely the procurement of material for offensive BW, and destroying existing stockpiles of BW agents and maintaining no stockpiles in the future.[30] The purpose of this memorandum was severalfold. First was the expression of support for the findings of the NSSM. Second was the strong suggestion that the president should make the announcement rapidly as the results of the NSSM were being "awaited by the press, the public and the Congress." Third, and perhaps the most interesting, the NSSM concluded that our capabilities "are weak in absolute terms, as well as relative to the Soviet Union."[31]

THE U.S. OFFENSIVE PROGRAM

The U.S. offensive BW program began in World War II. Initial efforts centered on understanding the potential of various pathogens to cause disease and identifying strategic or operational effects that could accrue from the successful deployment of biological weapons in times of war. These initial efforts, as with many of the state BW programs throughout history, have been shrouded in secrecy, often linked with special operations and clandestine national capabilities. This has made it difficult to gain a full appreciation of the U.S. program and the progress made toward developing BW capabilities.

The U.S. offensive BW program began in 1942 under the leadership of George W. Merck, who had been a member of the panel advising President Franklin D. Roosevelt on the war effort. The program was established at Camp Detrick, Maryland, with testing sites in Mississippi and Utah, and a production facility in Terre Haute, Indiana. The rationale for the program centered on investigating the potential for biological weapons to be used in a terror campaign much in the manner as the German high-explosive buzz bombs were used to terrorize England in 1943. The War Reserve Service (WRS) within the Army Chemical Warfare Service was to direct this effort. One account of the early effort states, "From the moment of its birth in the highest levels of government, the fledgling biological warfare effort was kept to an inner circle of knowledgeable persons. George W. Merck was a key and in May 1942 was charged with putting such an effort together."[32]

The program included research and development on a wide variety of biological pathogens including bacteria, viruses, and biologically derived toxins. The program was expanded during the Korean War (1950 to 1953), with a new production facility at Pine Bluff, Arkansas, that allowed for large-scale fermentation, concentration, storage, and weaponization of microorganisms; the actual production at this new facility was begun in 1954. The U.S. offensive program stimulated the complementary development in 1953 of a more robust defensive program to protect troops from a possible BW attack. The program included vaccines, antisera, and therapeutic countermeasures to protect troops from possible biological attack.

In 1952, the U.S. government began using live human test subjects in Operation White Coat during which more than 2,300 subjects, mostly Seventh-Day Adventists who had volunteered to serve in this manner rather than to fight in the war, agreed to be infected with live agents including *Francisella tularensis*, the bacteria that causes tularemia, and a purified toxin called *Staphylococcus enterotoxin* B (SEB). In these tests, medical counter-measures had to be given almost immediately after infection to prevent loss of life. While none of the human subjects were lost during the trials, the staphylococcal disease incubation times were far less than would have been predicted, leading to a better understanding of the concept of overwhelming dose. These tests also aided in the development of the concept of BW as having strategic capability.

U.S. cities were also used as experimental test venues, with BW simulants deployed in locations such as San Francisco, New York, and Washington, D.C. One account describes the program, "Cities were unwittingly used as laboratories to test aerosolization and dispersal methods; *Aspergillus fumigatus*, *B. subtilis* var. *globigii*, and *Serratia marcescens* were used as simulants and released during experiments in New York City, San Francisco, and other sites." In the San Francisco test, 130 gallons of liquid simulant *Serratia marcescens* was released off the coast of California in September 1950 and February 1951. The results were staggering, indicating that four hundred thousand people would have been infected. The cloud of simulant was effectively carried some thirty miles inland in infectious dose concentrations. The "harmless" release was later alleged to have caused an increased prevalence of urinary tract infections.[33] The infamous test of a light bulb filled with *Bacillus globigii* (the simulant for *Bacillus anthracis* or anthrax) was released in a New York subway in 1966 with extraordinary results.[34] The organism spread throughout the system within twenty minutes. In the Washington, D.C., test, a simulant was released in National Airport from several release points. Later the material was found to have "traveled" to several cities within a matter of hours.[35]

In the late 1950s and throughout the 1960s, small- and large-scale testing continued within the U.S. offensive BW program. In 1957 and 1958, a test called the Large Area Coverage or LAC was conducted to gauge the geographic dispersion and range of potentially released chemical and biological agents. The test involved the release of microscopic zinc cadmium sulfide (ZnCdS) particles over much of the United States and the tracking of these fluorescent zinc participles by ground stations. One account documents a four-hundred-mile release of five thousand pounds of zinc cadmium sulfide that was seen to have traveled over 1,200 miles to include from the release point into Canada.[36]

By the late 1950s, indications were that the potential for biological weapons were mixed. In a letter from one of Eisenhower's President Science Advisory Council (PSAC) members, George Kistiakowsky, to James Killian, the chair of the PSAC, BW was characterized as the following:

- Inexpensive methods have been developed for growing 10^9 organisms per cubic centimeter of liquid, suspending them, and storing them for months.

- Recent developments of freeze-drying techniques convert the liquids into dry powders with more than 10^9 live organisms per gram.
- Methods have been developed for producing aerosols from liquids that generate particles as small as five microns and contain live organisms.
- Limited viability of organisms in aerosols have been produced from liquid suspensions, with the "half-life" on the order of ten to one hundred minutes at night.
- Aerosols produced from freeze-dried organisms have a much greater viability and should be effective many hours after generation, even in moderately dry atmospheres.
- Infection by massive doses (e.g., 10^3 organisms) results in shorter incubation time and is able to overcome immunity by vaccinations.
- There has been an initiation of the selection and modification of strains or organisms to improve their resistance to antibiotics, change mortality rates, and more.[37]

Much had been learned over the years of the U.S. offensive program, yet concerns lingered about the potential to successfully use BW in a controlled and reliable manner. Many of these efforts centered on the methods of deployment (i.e., bomblets, cluster bombs, sprayers, etc.) and on perfecting these methods to avoid the loss of disease-causing pathogens in the process. The follow-up testing was designed to fill in these knowledge gaps.

To this end in 1962, President Kennedy authorized Project 112, which envisioned more than one hundred tests with names such as Night Train, Shady Grove, West Side, Red Cloud, and Speckled Start. The names of the tests were less important than the compilation of findings. We were learning more about BW, and these tests began to suggest potential scenarios in which the deployment of such weapons could provide a significant advantage. The tests had both strategic and tactical value for the U.S. offensive BW program. They allowed us to better understand the mechanisms of action of individual diseases. Which pathogens were the most effective? How did they cause disease? What were the most effective formulations for pathogens? Is it more effective to use liquid slurry or a dried powder formulation? What size particles were the most effective for the dissemination of aerosol clouds? How did encapsulation and stabilization affect the outcome of the deployment of a pathogen? How did meteorological and climate affect the pathogens? What was the biodecay rate of the various pathogens? What area coverage could be expected? What would be the effect if the pathogens were used together?

By 1963, military planners had developed enough interest in biological weapons that their use was contemplated during the Cuban Missile Crisis. Code-named "The Marshall Plan," the considered employment called for releasing incapacitating biological agents to attack defenders on the beach. In an interesting reversal of logic, the conclusion was that using BW would cause illnesses among the defenders and that they would believe they had to fight to the death rather than surrender. Therefore, the logic went that there would be an increase in U.S. casualties.[38]

In the 1965 Shady Grove test in the vicinity of Johnston Atoll, a liquid suspension of *Francisella tularensis*—the bacteria known to cause tularemia—was released in a line from spray tanks mounted to the wings of a U.S. Navy jet aircraft traveling at five hundred knots. In two minutes, thirty-five gallons of the *F. tularensis* was released over an unsuspecting group of rhesus monkey test subjects. The tests confirmed the presence of tularemia forty-five kilometers from the release point, and extrapolations indicate that a jet discharging 360 gallons could produce up to 50 percent human casualties over a 7,800-square-nautical-mile area. The results of what we were learning found their way into plans developed by the Joint Chiefs of Staff (JCS).

The military also experimented with combination weapons, drawing interesting conclusions. The pathogens could be deployed in combination so as to cause morbidity over a prolonged period of time. For example, an agent with a relatively short prodromal period could be used to cause illness for an initial period, at which time the second pathogen's prodromal period would end and the effects of the second pathogen were felt by the host. A cocktail combination was developed using tularemia and Venezuelan equine encephalitis (VEE) virus for achieving this sequential effect. Another way to combine biological weapons was to use them in combination to cause the pathogens to have synergistic effects. For example, using tularemia and SEB, one could achieve a greater lethal result on exposed personnel.

Officially, the U.S. military's assessment of BW from an Office of the Secretary of Defense 1967 study was that

> biological weapons were relatively poor weapons at their current state of development, that they could be developed into reliable, usable weapons, and that their development would be expensive, time-consuming, and of marginal impact on the overall strategic balance between the U.S. and the U.S.S.R.[39]

However, we also know that this assessment did not accurately represent the views of the JCS who felt that the weapons were further along in their development and did have "reliability and controllability."

During the NSSM 59 review, the military's position was that BW could serve as an important addition to its arsenal of weapons. However, BW was deemed to be an unlikely choice of weapon for first strike, but it could be used as part of a strategic campaign. There was even a contemplated doctrine for employment.[40] For example, one source noted the potential for: "preparing for an amphibious invasion, disrupting rear-echelon military operations or neutralizing pockets of enemy forces."[41] The NSSM noted somewhat counterintuitively that BW could actually in some cases be more humane than conventional weapons, as they could be used to incapacitate rather than kill the enemy. Despite the recognition of some of the serious flaws associated with the use of BW, the desire of the military was to continue to maintain offensive and defensive BW programs. Recognizing this to be untenable in the current political climate, the military's fallback position was to support the elimination of offensive BW, with the exception of toxins that were considered to be more akin to chemical weapons, and to continue with the defensive BW program.

In the NSSM deliberations, Staff General Earle Wheeler, the Chairman of the Joint Chiefs, was isolated in his position, which called for maintaining a full, offensive CBW capability. "With regard to our biological warfare program," Wheeler said, "its major value is deterrence. If this fails, then we have a modest ability to retaliate. . . . The JCS believes that, on balance, it has a low cost, that it would be a catastrophe if we can't respond, and there is a difficulty in verifying enemy capabilities. Therefore, the JCS believes that we must retain our present stockpile and the option of production if needed."[42]

Historically, few public pronouncements or declarations have been made by the United States with regard to offensive BW capabilities. A few data points provide only limited insights into U.S. thinking on the potential incorporation of BW into our arsenal. President Roosevelt's statement about no first use of poisonous gas but having a capability to retaliate in kind provides the foundation for a "no first use" doctrine. Policy guidance from 1959 Basic National Security Policy (BNSP) discusses the potential for using biological weapons, if necessary, to enhance the effectiveness of the Armed Forces. In a 1966 letter from the Secretary of Defense to the Secretary of State, the Department of Defense requirements for maintaining biological weapons was linked to having a balanced offensive and defensive capability and for possessing them as a deterrent against an enemy.[43]

The U.S. offensive program under the auspices of Project 112 had developed several important understandings of the potential of BW. The weapons, when properly prepared and deployed, could have strategic effects. The capabilities of BW, once considered to be too difficult to control, could actually be made to have predictable effects. There really was no such BW weapon as an incapacitating agent. At the appropriate levels, any biological pathogen could cause high mortality and morbidity. Finally, BW weapons did not behave as naturally occurring disease.

THE SOVIET OFFENSIVE BW PROGRAM

The NSSM provided some conclusions about the Soviet Union's BW capabilities that were important to the U.S. unilateral decision to renounce BW.

The Soviets were assessed to have the capacity to develop offensive biological weapons, but they noted that their efforts were tightly controlled with a "high degree of political control and restraint."[44] The NSSM 59 conclusion that our capabilities "are weak in absolute terms, as well as relative to the Soviet Union" is both counterintuitive and not in accordance with the Soviet view of the issue. The United States had over the twenty years leading up to 1969 concluded a series of tests that had established that biological weapons had strategic capabilities. The Soviets—we would later find out from various defectors' reports—began their massive biological weapons program after the BWC entered into force in 1975.

Arguably, the most robust of all state programs was that of the Soviet Union, which continued in earnest from the 1930s until the early 1990s. The early years of the program were primarily focused on laboratory work to understand the potential for the use of pathogens in conflict. The significant buildup of the Soviet program began with the establishment of Biopreparat in 1973 as a "civilian" follow-up effort of the earlier military Soviet biowarfare programs. Much of what was learned about the program came from Soviet defectors. In 1989, Vladimir Artemovich Pasechnik, who had served as a senior Soviet scientist within the offensive BW program, defected to the United Kingdom. In a series of debriefings, Pasechnik began to describe an offensive BW program of a scale and scope that far exceeded anything that we had previously considered, and certainly far greater by an order of magnitude, than the U.S. program. Where the United States had conducted tests and developed stockpiles of key biological agents, the Soviets had created an industrial capacity for producing, storing, loading munitions, and disseminating biological weapons of mass destruction. The Soviets had progressed so far in their program that they had developed the doctrine of use for striking at strategic targets, including the U.S. homeland, using intercontinental ballistic missiles.

Pasechnik's revelations were confirmed in 1992 with the defection to the United States of Colonel Kanatjan Alibekov, the number-two scientist for the program. Alibekov, who later Americanized his name to Ken Alibek, served as the deputy for Biopreparat, the organization responsible for leading much of the Soviet BW development program.[45] The Soviet program included all aspects of BW weapons design from the laboratory work to the open-air testing to the development of the delivery systems that would transport the deadly cargo thousands of miles to their intended targets; the Soviets reportedly had developed the capability to launch BW-capable intercontinental ballistic missiles. For ensuring the availability and viability of the weapons, the Soviets had factories for BW production at Stepnogorsk in Kazakhstan and six other facilities that could rapidly produce large amounts of the weaponized BW material from the seed stocks that they maintained.

The Soviets melded their civilian and military biotechnical capabilities into a highly secretive program that included hundreds of facilities and thousands of scientists. The often-cited Vozrozhdeniye (Rebirth) Island in the Aral Sea was the prototypical open-air testing facility that even today after closure and significant decontamination at the expense of the United States as part of the Nunn-Lugar Cooperative Threat Reduction Act remains a significant health hazard. Mortality rates and health statistics from the region indicate significant anomalies, such as high levels of arsenic that are continuing to lead in large numbers to low life expectancy and high infant mortality, despite billions of dollars invested in clean-up efforts.

One noteworthy event provides evidence of the program the Soviets had developed, the secrecy surrounding the program, and the potential lethality of the weapons. In 1979, an accidental release of anthrax occurred from a weapons production facility in the Soviet Union in the city of Sverdlovsk. The outbreak, which occurred over a sixty-day period, was originally publicly attributed to contaminated meat, but in the age of Perestroika, President Boris Yeltsin acknowledged the accidental release was from a weapons production facility. The incident has also been well documented in the Alibek book. The release was apparently caused by the inadvertent removal of a filter that was used in the end-stage production of the bacterium spores into appropriately sized particles for aerosolizing through the process of milling. The filter was essentially the last line of defense preventing the weaponized spores from being released to the outside air. The release formed a highly predictable plume and subsequent dispersion pattern based on meteorological data at the time. The "contaminated meat" caused ninety-six people to become ill and sixty-six to die, according to the accounts in the Alibek book. The length of time from the release of the spores to the final infections provides evidence of the hardiness of the anthrax spores and the ability of this pathogen to cause infections. While the initial release created a plume in the atmosphere and sickened many of the night-shift workers on duty within the affected area, later infections were caused by the spores being reaerosolized in the days and weeks after the original release. It is interesting to note that one assessment states that the release "exposed 15,000 workers at the plant and at least 50,000 people living in the surrounding area to an aerosol of anthrax spores."[46]

The experimentation by the Soviet Union included efforts to develop resistant forms of disease and to increase the lethality of their pathogens. One account in which an antibiotic strain of anthrax was examined is provided below:

> Scientists found that a plasmid with the genes for resistance to tetracycline, one of the most potent and widely effective of all antibiotics. In a Petri dish, they mixed small quantities of *B. thuringiensis* with anthrax, cultivated the two

strains together, and then placed them in a test tube with tetracycline to see if the anthrax bacteria would survive. The antibiotic killed most of the anthrax bacilli, but a few cells survived. Most of these had incorporated the antibiotic-resistant genes from *B. thuringiensis* into their own genetic structure. These new cloned cells could now be used to create tetracycline-resistant strains of anthrax.[47]

The program began to be dismantled in the early 1990s, yet remnants remain today in various states of disrepair in a number of countries formerly part of the Soviet Union. Work continues under the Nunn-Lugar program to destroy the last vestiges and to ensure even the knowledge gained through the program is no longer used for the development of BW.[48] The motivation for the massive Soviet build-up even today remains unclear, although one cannot argue against the veracity of their significant biological weapons development program.

OTHER OFFENSIVE BW PROGRAMS

In examining the potential for biological weapons to be used by other countries, NSSM 59 concluded that even countries with limited industrial bases were likely to have or be able to develop the capacity to use biological weapons. The NSSM did not identify any other specific national offensive BW programs but noted the potential for BW to potentially provide a strategic capacity to a nation that was approaching nuclear capability at a small fraction of the cost.

This belief was reinforced by the history of offensive BW exploration up to that point, especially during the interwar years. During this period, several nations developed offensive BW programs. In Europe, the United Kingdom, France, and Hungary had such offensive BW programs. The German military scientific experts had discussed the possibility of biological warfare in the 1920s, but they determined that BW was not "practical as a method of warfare." Despite this German assessment, strong suspicions existed that the Germans used BW directed against livestock in World War I to disrupt Allied logistics.

Outside of Europe, Japan developed a significant BW capability that included development, testing, and even allegations of use in warfare. Japanese General Ishii Shiro, an army medical doctor, ran Unit 731 from 1931 until 1945. The Japanese efforts were clearly the most advanced of any state BW program at the time. The program grew to over fifteen thousand people and was infamous for human experimentation including the use of BW against Chinese citizens and enemy prisoners of war at a cost of thousands of dead. At the Ping Fan facility located south of the city of Harbin, the Japanese were able to produce kilogram quantities of bacteria that cause plague, anthrax,

typhoid, cholera, dysentery, and other diseases.[49] The Japanese also conducted large-scale deployment of biological weapons on several occasions.
In 1939, the Japanese allegedly poisoned Soviet water sources with intestinal
typhoid bacteria. Then in 1940, Unit 731 dropped paper bags filled with
plague infested fleas on two Chinese cities.

Finally, in September 1945, the Soviet Army overran Unit 731; many of
the facilities and records were destroyed in the process. Some of the Unit 731
leaders were captured and tried by the Soviets, however, the U.S. dismissed
these trials as "propoganda". In the post war autopsy of over 1,000 victims, it
was revealed that most had been exposed to aerosolized anthrax. General
Ishii was later captured by U.S. forces and several of the leaders of the unit
later struck a deal to provide details of the work at Unit 731 in exchange for
immunity against being charged with war crimes. Both the United States and
Soviet Union conducted interviews with Unit 731 members and used the
information for increasing understanding of biological warfare, ultimately
for developing their respective offensive programs in the post-World War II
aftermath.

To be balanced in developing the potential for use of offensive BW, one
area that has tended to separate BW from even nuclear and chemical weapons has been in the use of these weapons for deterrence. In this regard, BW
has never served as a deterrent as have nuclear and chemical weapons. While
nations have through declaratory policies and even through use in the case of
chemical weapons discussed the potential use of these capabilities, we have
seen no such declarations with regard to biological weapons. This is both
curious and reasonable, given societies' general concerns about the deliberate use of disease as a tool of warfare. The public relations and humanitarian
aspects of a declaration of intent to use biological weapons likely have limited the appetite for using a biological weapon deterrence framework.[50] In the
modern era of the BWC, the only arguable exception to the use of BW as a
deterrent can be made with regard to Iraq's Saddam Hussein, who at best was
deliberately ambiguous and at worst used his stocks of BW as a deterrent in
the Iran-Iraq War.

One account describes the Germans' view of the use of biological pathogens in World War I to cause disease in the following manner, "[The] use
[of] biological weapons to kill or injure humans was immoral, but that biological attacks on horses and other draft animals were entirely legitimate."[51]
In explaining this lack of deterrent value, one author introduces the concept
of "dreaded risks," a category of risk in which the fear of an act is disproportionate to the actual outcome, and places BW in this category. To demonstrate this concept, the author cites the 2001 anthrax attacks in the United
States in which five people died. The response, resulting in spending billions
of dollars and establishing new defensive programs and capabilities, has
elicited a response well beyond what seems reasonable given the demonstrat-

ed threat; one author maintains this is due to this concept of dreaded risk of biological weapons.[52] This reaction stems from several important, interrelated factors. Horror of the disease is based on the inherent fear of disease and possible contagion. The loss of faith in the ability of scientists to protect pertains to both the idea of inevitability and the relatively new concept that science can be used for nefarious purposes, which diminishes our "technological optimism."[53] This framework should be both heartening and of some concern to those looking to prevent a terrorist BW attack as it serves to identify potential limits or thresholds on the use of BW as well as signify that in some cases, the terrorist may find the use of BW to be an acceptable means for achieving an intended outcome.

In 1995, the Office of Technology Assessment at Senate hearings listed seventeen countries that were "suspected of manufacturing biological weapons."[54] While this is well after the BWC entered into force, it provides some perspective on the relatively small numbers of nations that have had BW programs. This number reflected an estimate rather than a definitive assessment of the number of national programs. The question of the definitive number of BW-capable states was then and remains now a difficult one. A number of nations are thought to have had programs that included the acquisition of pathogens and basic research. The research in most cases included processing the pathogens to alter basic characteristics. This included conferring antibiotic resistance or even using synthetic biology techniques. However, very few of the nations were thought to have gone beyond the early steps of acquiring and processing the pathogens. Even fewer had developed a weaponized capability that would include the pathogen and the means of deployment. In short, the number of nations that are thought to have fully weaponized and stockpiled quantities for large-scale use is quite small. Included in the totals were a number of nations that had developed small BW programs more in keeping with assassination purposes rather than for large-scale deployment.

THE UNITED STATES' DECISION ON OFFENSIVE BW

Several important findings resulted from the NSSM 59 study. The United States needed to retain a research, development, test, and evaluation (RDT& E) capacity to improve defense and guard against technological surprise. Increased control mechanisms needed to be put in place for working with these elements. Better intelligence would be required for understanding other nations' efforts. Our U.S. declaratory policy should be "no first use." No agents, with the exception of riot control agents, should be used without presidential approval. Finally, a public relations campaign was required to align the perceptions with the reality of the program.

The findings of NSSM 59 were also colored by the public attitudes toward chemical and biological weapons. Communist "propaganda" concerning the U.S. failure to ratify the 1925 Geneva Protocol, the U.S. use of riot control agents in Vietnam, and the continued claims that the United States used germ warfare in Korea in the 1950s put the United States on the defensive with regard to biological weapons.

During the final stages of the NSSM 59 deliberations, positions became hardened. Secretary Laird argued for the decoupling of chemical and biological weapons as they were "entirely different subjects." He also noted the important lack of a "deterrent quality" concerning BW. Laird also stated clearly his belief that the United States should renounce offensive biological warfare but retain a defensive research and development program to develop vaccines and other medical countermeasures. Secretary of State William Rogers and Arms Control and Disarmament Agency (ACDA) director Gerard Smith essentially supported Secretary Laird's stated position. At this point in the deliberations, General Wheeler was isolated and decided to back down, joining the growing consensus in favor of eliminating the offensive BW program. "We don't feel as strongly about BW as about CW," he explained. "We would like to see a minimal RDT&E [research, development, testing, and evaluation] program pointed to defense, guarding against offensive actions by the enemy."[55] At the same time, Wheeler took a hard line with respect to retaining U.S. offensive CW capabilities for deterrence and retaliation, as well as the continued combat use of tear gas and herbicides in Vietnam. Given the consensus of the interagency principals, President Nixon also expressed support to eliminate the offensive component of the U.S. program and to confine the BW program to defensive R&D only.

President Nixon's simple and clear statement during a press conference in the Roosevelt Room of the White House on November 25, 1969, seemingly closed the chapter on the U.S. offensive BW program. He stated, "Biological weapons have massive, unpredictable, and potentially uncontrollable consequences. They may produce global epidemics and impair the health of future generations."[56] Almost immediately after the statement, the president was forced to confront a growing confusion and potential loophole caused by the ambiguity concerning biologically derived toxins that were technically excluded from the renunciation. This omission was corrected through NSSM–85, "Review of Toxins Policy," which was issued on December 31, 1969. The NSSM allotted two weeks for the study, which was justified by the narrow scope of the topic, coupled by the fact that State, Defense, and the JCS were already preparing an options paper on toxins and the growing controversy. After an abbreviated and necessarily timely review, the White House issued a clarification on February 14, 1970, that expanded the original renunciation to include "offensive preparations for and the use of toxins as a method of warfare."[57]

It was within this context that National Security Decisions 35 and 44 were signed by President Nixon in November 1969 (microorganisms) and February 1970 (toxins), respectively. These decisions mandated the cessation of offensive biological research and production and the destruction of the biological arsenal. Any further research efforts were directed to the development of defensive measures such as diagnostic tests, vaccines, and therapies for potential biological weapons threats. The U.S. stocks of pathogens and the entire biological arsenal were destroyed between May 1971 and February 1973 under the auspices of the U.S. Department of Agriculture, the Department of Health, Education, and Welfare, and the Departments of Natural Resources of Arkansas, Colorado, and Maryland. Fort Detrick in Frederick, Maryland, was permitted to retain small quantities of pathogens to test the efficacy of investigational preventive measures and therapies. The inclusion of the ban to cover "all toxins, regardless of their means of production, closed the potential loophole that would have been created by the future chemical synthesis of toxin agents and resulted in a U.S. policy that was cleaner and less ambiguous."[58]

The U.S. unilateral statement to renounce biological and toxin weapons auspiciously became the first time that a major power unilaterally abandoned an entire category of armament. Further, it cleared the way for the rapid negotiation of the 1972 Biological and Toxin Weapons Convention (BWC), which banned the development, production, stockpiling, and transfer of biological material that was for other than peaceful, preventive, or prophylactic purposes. In 1974, President Gerald R. Ford submitted the BWC and the 1925 Geneva Protocol simultaneously to the U.S. Senate for its advice and consent for ratification. Consent was duly granted on December 16, 1974, and the BWC entered into force on March 26, 1975.

The BWC became the first arms control treaty that eliminated an entire class of offensive weapons. Since entry into force, the number of member states parties has grown from 22 to 168 nations, with several others as only signatories. While the numbers have grown since entry into force, the BWC still lag behind similar treaties such as the Chemical Weapons Convention (CWC) and the Nuclear Nonproliferation Treaty (NPT) with 188 and 190 member nations, respectively.

Chapter Two

Biotechnology and its Implications for the BWC

In the 1800s, we saw huge changes from the invention of the steam engine and the ability to create and distribute energy. In the 1900s, we saw revolutionary changes from electronics and the ability to manipulate information. The major change in this century will be biological control over manufacturing. From an engineering perspective, biology is about making things—it's a manufacturing technology. It puts atoms precisely where they should be. [1]
—*The Industrialization of Biology and Its Impact on National Security*
University of Pittsburg Medical Center

We have been discussing biotechnology and alluding to the important changes that are occurring with breakneck and at times unchecked speed. However, we have been discussing the issue in a somewhat superficial manner. Given the centrality of the biotechnological revolution to the premise of this book, a more detailed account of the ongoing change in the field is essential. Further, the importance that many nations place on the discussion of the topic within the context of the BWC review conferences and other sponsored events indicates the benefit from a more in-depth review of the topic.

We will examine biotechnology from several unique perspectives. First, we will delve more deeply into the changes that are occurring and see how they impact not just on our daily lives but also the potential for misuse, both inadvertently and deliberately. We will also look at the convergence of several different fields of study that are combining and converging with extraordinary speed and highly impactful results that make this the Age of Biotechnology. Finally, we will consider the subject from an orthogonal perspective considering how and why the BWC evolved into its current form.

To frame the debate on biotechnology within the BWC, the perception is that developed nations are seen to have a large advance in biotechnology, and developing nations are seeking to gain access to advanced technologies under Article X of the Convention. Of course, the biotechnology issue is not as simple as a sharing of national capabilities. In fact, nations are not the exclusive holders of capabilities in biotechnology; rather, most biotechnological capacity is contained within private or publically traded companies to include multinational companies that have economic incentive for development. Laws and regulations including intellectual property provisions and export controls govern significant aspects of biotechnology. One common issue that surfaces at every BWC event from the intersessional process to the review conferences is the question of the continued relevance of the BWC in light of the changes that have occurred since the BWC entry into force (EIF). To date, the collective parties have agreed that the BWC provisions remain relevant despite these advances.

THE BIOTECHNOLOGY PROBLEM

No single authoritative definition of biotechnology exists. Early definitions centered on the manipulation of the human genome, which is often referred to as DNA engineering, recombinant DNA technology, or genetic engineering. While these descriptions remain relevant today, they are incomplete and do not fully account for the changes occurring in biotechnology, our understanding of the field, and the growing ability to manipulate biological material. In short, we are experiencing a revolution in biotechnology that will fundamentally alter the possibilities for life on the planet. By all indications, we are in the early or at least earlier stages of the metamorphosis, with some suggesting that the most significant understandings are yet to be discovered.

It is a commentary on the state of biotechnology today that far more was understood about the workings of gravity than the inner workings of a cell. We had put a man on the moon before we had been able to prove that the DNA composition of humans is 99 percent similar to chimpanzees and gorillas in 1972. Our knowledge of biology has progressed rather slowly as compared to other fields such as physics, material sciences, and even information technology.

Humans have used biological processes for millennia, yet we did not understand the basic science behind such biological processes as the brewing of beer, which dates back to the Sumerians in 1750 BC. The first microscope was invented by Zacharias Janssen in the 1590s, cells were first described by Robert Hooke in 1663, Antony van Leeuwenhoek discovered protozoa and bacteria in 1675, James Watson and Francis Crick published their paper describing the double helical structure of DNA in 1953, yet it took until 1999 to decipher the genetic code of the complete human genome. It took only three years from this milestone for researchers at the State University of New York at Stony Brook to recreate *de novo* the poliovirus and eleven years for researchers at the J. Craig Venter Institute to create the first synthetic, self-replicating cell.

From the linear progression, some might even say plodding growth, of early biology to the exponential expansion we have been experiencing over the past couple of decades, it is no understatement to claim that we are in an explosion in the field of biotechnology. Still, the progression has not been uniform across the endeavor, which is hindering overall progress in biotechnology. For example, great strides have been made in the area of biosequencing, which is the reading of the genetic code of an organism. Yet much less progress has been made in bioinformatics or the making sense of the volume of data that has become available from sequencing. We are able to decode the blueprint but have little ability to correlate and understand how variances in genetic code translate to differences between the traits of organisms or variances in the virulence of pathogens of the same species. To date, gene therapy has shown great promise, but it has been most useful for manipulating individual traits in plants and animals and far less than advertised for widespread manipulation of genetic sequences of more complex species.

By way of an analogy, our understanding of the composition of the human genome is very much like having a map with all of the cities on it yet none of the roads or directions for transiting between cities. We have decoded the approximately 3.3 billion base pairs of DNA; laid out the organization of the twenty-three human chromosomes into base pairs, proteins, messenger ribonucleic acid (RNA), protein coding genes, simple sequences, and the like; and know that there are some twenty thousand protein-coding genes; yet we have only cursory understanding of the critical signaling pathways that govern even the simplest of interactions within the human species, or any other species for that matter.

As an example of this overall lack of understanding of the complexity of DNA structures and interactions, consider a project called ENCODE (Encyclopedia of DNA Elements). The project that included over 440 scientists at thirty-two laboratories around the world examined the human genome in greater detail to gain an understanding of how the DNA sequences actually interacted. ENCODE identified over four million gene switches that reside in bits of DNA that once were dismissed as "junk" but that turn out to play critical roles in controlling how cells, organs, and other tissues behave. The ENCODE project provides information on the human genome beyond that contained within the DNA sequence, describing "the functional genomic elements that orchestrate the development and function of a human."[2]

Specifically, the project examined the "junk"—parts of the DNA that are not actual genes containing instructions for proteins—and discovered that genes are controlled by a complex system that assists in the regulation of the cells and in combination determines the uniqueness of each individual. They also determined that some 80 percent or more of this DNA is active and needed, focusing on the signaling biochemical functions. The result of the work is an annotated road map of much of this DNA, noting what it is doing and how. The system includes a series of "switches that, acting like dimmer switches for lights, control which genes are used in a cell and when they are used, and determine, for instance, whether a cell becomes a liver cell or a neuron." The study also concludes that even in cases of identical twins, "small changes in environmental exposure can slightly alter gene switches, with the result that one twin gets a disease and the other does not."[3]

It is this "junk DNA," or the space between genes, that is "filled with enhancers (regulatory DNA elements), promoters (the sites at which DNA's transcription into RNA is initiated), and numerous other previously overlooked regions that encode RNA transcripts that are not translated into proteins but might have regulatory roles."[4] In other words, the gene sequences are the blueprint for the life-form, while the proteins serve as the electrical connections and turn functions on and off.

The significance of ENCODE is such that it bears repeating. In discussing biological process and life, the assumption has been that the DNA material or code has been the determinant of the traits of a life-form. Now we are beginning to understand that while DNA provides an important blueprint for this description, underlying switches are the final determinants of whether disease will occur. It also signals that while genes are essential to describing a life-form, the combination of DNA switches serve as the underlying instructions for determining how the organism will react. This finding, only possible through bioinformatics, will undoubtedly lead to changes in fighting disease and delivering on the promise of personalized medicine. It also begs the question of what other underlying truths are left to be discovered.

A convergence is occurring within and between several fields, which is demonstrating that the whole is far greater than the sums of the parts. These advances and the potential have not gone unnoticed. Nations have staked their economic futures on biotechnology. In several Asian countries such as China, India, Malaysia, and Singapore, biotechnology has become a key element of national economic development plans.[5] Even potentially dangerous pathogens discovered within a nation's borders are being considered for their economic value; what if a pathogen is used in the development of a vaccine or is therapeutic? The source nation hopes to be able to share in the proceeds that are generated. Such was the case with the U.S. joint laboratory in Indonesia that was closed, in part due to differences in the manner in which the Indonesians and the United States saw the potential compensation for a strain of influenza.

Within the field of biology, extraordinary advances are occurring. Consider one account:

> [Rapid] advances in mapping the human genome (genomics), studying the structure and function of the myriad of proteins in living organisms (proteomics), and analyzing the complex bio-chemical circuits that regulate cellular metabolism (systems biology) are yielding a profound new understanding of life at the molecular level. [6]

The account goes on to describe the convergence of biological and chemical production methods that are being used interchangeably in the production of drugs and chemicals. What does this portend for the continued separation of the biological and chemical arms control fields in the BWC and CWC, respectively? Is there a distinction any longer, or have technological advances (and those anticipated into the future) created a logical convergence of the two conventions?

Many experts combine several different fields in developing definitions, or perhaps what might be better termed as explanations, of biotechnology. We have already alluded to the symbiotic relationship between biology and chemistry; consider what happens when these fields are combined with nanotechnology, information technology, advanced manufacturing, and neuropharmacology. These multidisciplinary, complex interactions combine to create an acceleration of capacity in which the rising tides lift all boats. Having advanced information technology—which is also referred to as bioinformatics—allows for rapidly comparing DNA sequences, which allows for faster drug development and a greater understanding of the interactions of the therapeutics. Biological processes can be used in the development of chemicals. Biocatalysts, such as enzymes, can be used to synthesize chemicals, and likewise, chemicals such as polymer-grade acrylamide can be produced using biotechnology; this would allow for "chemicals" to be produced more efficiently and more environmentally friendly for applications such as cleaning compounds and detergents. [7] Advanced manufacturing techniques such as nanotechnology, which is related to three-dimensional (or 3-D) printing, allow for the printing of genetic material based on the rapid ability to describe and copy genetic sequences. One forecast states that by 2015, 20 percent of the U.S. $1.8 trillion chemical industry could be dependent on synthetic biology. [8]

Let's consider the 3-D printing example in more detail. Genetic material or DNA (deoxyribonucleic acid) consists of a combination of four nucleotides—adenine (A), guanine (G), thymine (T), and cytosine (C)—combined with sugar and phosphate. The combination of material forms the DNA double-helix shape where A and T always pair and G and C always pair to provide a biological blueprint that gives an organism a specific trait or characteristic. This blueprint describes for the cell how to produce the specific proteins or gene sequences for the over ten thousand different gene sequences found in most plants and animals and approximately twenty-five thousand gene sequences in mammals. In turn, the set of genes for an organism are organized within the cell nucleus into chromosomes, and it is this set of chromosomes that combines to provide the unique DNA makeup and combination of characteristics and traits that combine to contribute to the uniqueness of every living organism. Through the rapid ability to sequence— that is, to identify the DNA-defined structure of an organism—we are now able to quickly and efficiently define all forms of life. Sequencing requires scientists cut the DNA up into short fragments, sequence the fragments simultaneously, and assemble the entire genome by using sophisticated computer techniques to match the fragments to each other. Notice that sequencing is highly dependent on high-speed computing resulting from advances in information technology. Advanced manufacturing techniques are now available that allow for the development of various structures using the raw material and a computer-aided design (CAD) process based on X-ray crystallography to describe the structure. So if the genetic blueprint of an organism is known, imagine the potential for using a four-head printer that uses the proteins A, T, G, and C as outputs. The CAD portion of the process would be needed to ensure that the proteins are folded correctly and that they would function as the original organism. This 3-D printing example is highly complex and certainly at the edge of today's technology, but as we have witnessed over the course of human history, once the science has been discovered, the engineering of the science can be accomplished given the appropriate motivation and resources. As one author notes, "Many areas of biotechnology are progressing at an exponential rate, and the lag time from scientific discovery to technological application is extremely short."[9]

We have already seen the direct application of this synthetic biology in areas such as genetically modified organisms (GMO) in agriculture. The insertion of a gene, Bt, to provide protection to crops such as corn, cotton, and potatoes against insects allows for the use of less pesticides and targeted protection. For example, Bt gene insertion provides corn that is resistant to the European corn borer.[10] Other GMO applications can be incorporated to improve yield, increase drought resistance, and produce insect resistance. Additionally, GMO can also be incorporated into the agricultural sector to improve food safety and security to decrease the time in detecting food pathogens, toxins, and chemical contaminants. Limited GMO usage has become mainstream, yet will undoubtedly expand as the technology matures and populations become more comfortable with the concept of genetically modified organisms in the human food supply chain. Globally, different levels of support have been expressed for GMOs, with the loudest opposition coming from Europe, cautious support from the United States, and greater support among many developing nations, particularly in Africa and Asia that see the potential benefits resulting from higher crop yields and less endemic disease.

Other applications are employed today that are directly used for human consumption. Insulin is produced using a biochemical process whereby a harmless strain of *Escherichia coli* bacteria is given a copy of the gene for human insulin. Gene therapy is being incorporated into the toolset for fighting diseases including cancer, cystic fibrosis, and AIDS. Malnutrition in developing nations is being addressed using "golden rice" that is genetically modified to provide higher levels of Vitamin A. Efforts are underway to develop crops with higher iron levels. We are in the early stages of these efforts. As our knowledge continues to expand and other technologies are incorporated in a synergistic manner, the future will continue to see advances in both the knowledge and practical application of biotechnology for the benefit of humankind.

The future for biotechnology is bright with the promise of great opportunities for changing our daily lives in dramatic ways. Biomedicine demonstrates the potential for complex molecular capacities for tissue repair, smart therapeutics, novel delivery systems for disease-fighting, personalized genomic medicine, and *in vivo* drug synthesis by microbiome manipulation. For the environment and energy, we are already seeing the potential for bioremediation from oil-eating bacteria, "green" energy microbial production, and the potential for safe GMOs. Other areas for likely growth include *in vitro* synthesis of complex biopharmaceuticals, a sustainable "green" chemical (microbial) industry, and the production of smart (bio) materials.

However, these changes in biotechnology magnified by the changes in synergistic fields also provide a window into the potential for the same technology to be used for inadvertent or nefarious purposes to great effect. As we continue to feel the effects of the synergies between biotechnology and information technology, we will gain greater knowledge into the biological sciences. Today the issue is that there is too much data and not enough capacity for manipulating, visualizing, and analyzing it. Once we are able to understand the vast amounts of data that are resulting from techniques such as genetic sequencing, we will begin to realize the potential of the Age of Biotechnology.

Finally, this explosion in knowledge within the field has been found to be inversely proportional to the cost implications of developing these biotechnologies. While it cost some $2.7 billion to complete the human genome project, estimates are that in 2014 the cost of sequencing an entire human genome will have dropped to $1,000. Further, the accessibility of the technology and the convergence of multiple related fields will fundamentally alter how biotechnology affects humankind.[11]

THE INDUSTRIALIZATION OF BIOTECHNOLOGY

In describing the changes underway in the field of biotechnology, one source states, "Almost anybody nowadays can pick up textbooks on methods or read current papers on methods and techniques that 10 years ago were the stuff of Nobel prizes. The knowledge is becoming commoditized."[12] Another author sums up the changes in biotechnology as "becoming increasingly de-skilled and less expensive."[13]

This commoditization surely will provide new opportunities for economic development and benefits to humankind, yet the potential also exists for the lowered thresholds to place greater capabilities into the hands of those who would seek to misuse the technology, either intentionally or inadvertently. And unfortunately, this proliferation of capabilities does not come with an on/off switch or the necessary protections to ensure that it is only available to worthy individuals.

We see these trends in processes that once required great skill but have now become industrialized. The growth of biological material is no longer an art but rather a well-understood science. Follow the directions on the pre-packaged kit and the bacteria or virus will grow. Experts no longer have to understand the intricacies of developing growth media for culturing. Likewise, processes such as polymerase chain reaction (PCR) have been automated and will be done "automatically" in a matter of thirty minutes or so. The author that describes the "deskilling" of the process uses the term to imply a reduction in the amount of tacit knowledge that one must have to be able to achieve a desired outcome.[14]

In previous state offensive BW programs, experts with highly specific knowledge were required to conduct even the basic steps inherent in the development of biological weapons. With the infusion of engineering processes, the tacit knowledge required has been reduced to a small number of steps involving a subset of the original tasks. For example, if there were ten complex tasks associated with the development of a BW weapon and half of the steps have been reduced to an engineered solution requiring little specific knowledge, then a bioweaponeer is far closer to achieving such an offensive capability. As the field of biotechnology continues to mature, we should expect that thresholds for developing BW capabilities will continue to be lowered. This implies that more people will have access to these building-block steps and therefore that entry into what heretofore has been the exclusive territory of sophisticated states will be available to a wider array of states *and* people, some with less-than-noble intent. The degree to which this "democratization of biology"[15] has occurred can be seen in the various organizations and competitions that promote the proliferation of biotechnology, including the Do-It-Yourself (DIY) movement and the Internationally Genetically Engineered Machine (iGEM) that is held at the Massachusetts

Institute of Technology (MIT) by the BioBricks Foundation.[16] The competition has grown from five teams in 2004 to 165 teams in 2011.[17]

Community laboratories have been established where space can be rented, just as one would do in getting a hotel for the night. This increasing access has significant potential for misuse by amateur biologists and lone-wolf actors, as well as for terrorist organizations seeking to promote their causes. One author cautions that synthetic biology could be deskilled to the point where amateur scientists, hobbyists, and even children, will use it to play "biotech games, designed like computer games for children down to the kindergarten age, but played with real eggs and seeds rather than with images on a screen."[18] Even such mundane issues as the appropriate destruction of DIY waste have not received the scrutiny they deserve. Imagine the potential for the remnants of a DIY experiment to be ingested by an animal and potentially cause a disease outbreak.

The fruits of the biotechnological advances are becoming more widespread and cost effective. Using a technique of inserting a virus into another virus, doctors are able to treat cancer patients using an adenovirus and alphavirus combination to target certain types of cancers. This approach shows great promise for the treatment of other types of disease as well. This technique for the treatment of cancer also has much in common with the BioBricks discussion above. BioBricks are standard biological DNA sequences that have defined structures and functions. They have been constructed to have a common interface and are designed to be composed and incorporated into living cells such as *E. coli* to construct new biological systems. In considering BioBricks, think about a Lego set that allows one to construct various structures. Just as experts were able to use a targeted virus within virus methods to treat cancer, a member of the BioBricks community could develop combinations of biological material, some of which could have lethal consequences. Consider the delivery of a deadly virus such as rabies within a "harmless," nonpathogenic *E. coli* delivery system. This is not to imply that members of the BioBricks community have ill intentions. However, a deadly combination of BioBricks could just as easily be created through experimentation gone awry.

By way of an example of the inherent difficulty in developing predictable synthetic biology outcomes, consider the story of the retrovirus Xenotropic Murine Leukemia Virus-Related Virus (XMRV), which was produced inadvertently in the laboratory in the 1990s by the passage of a prostate tumor in mice. Although XMRV is not known to cause human disease or is not known whether or not it has the potential to do so, the virus can infect a variety of cultured human cells including peripheral blood mononuclear cells and neuronal cells.[19] The previously discussed mousepox experiment, in which the IL-4 gene was introduced and made the virus far more lethal, is another example of the difficulty in predicting synthetic biology outcomes.

This discussion of synthetic biology has important applicability to both the BWC and the potential for misuse or use of biological material for weapons. One of the important steps in the development of a BW capability is the acquisition of the biological material. Conventional wisdom had been that the primary threat for acquisition would be to get the pathogen from a natural source. With the synthetic biology progress, we have seen the increasing potential for the development of a pathogen. Consider the advances that have been made. In 2002, the first functional virus was synthesized. It had approximately 7,500 nucleotide base pairs. In 2004, techniques had been improved that allowed for synthesizing sequences of 14,600 and 32,000 nucleotides. In 2005, scientists had been able to reconstruct the 1918 influenza virus. By 2008, the Severe Acute Respiratory Syndrome (SARS) virus had been re-created, and the J. Craig Venter Institute had synthesized the genome of a bacterium consisting of 583,000 base pairs. By 2010, they had been able to synthesize the entire genome of the *Mycoplasma mycoides* bacterium consisting of over one million base pairs and its replication in bacterial cells.[20]

It is an understatement that great progress has been made in synthetic biology. Yet as the synthetic biology time line presented above indicates, we are continuing to see progress measured in exponential leaps, indicating that we are far from reaching the theoretical limits in this field. This discussion is not meant to imply that a terrorist will have the capability now or in the near future to *de novo* synthesize a biological pathogen, but rather that important thresholds are being lowered in synthetic biology that must be factored into our thinking. It also highlights the tenuous balance in the life sciences community: synthetic biology can and will lead to important benefits for humankind and therefore, we want to allow legitimate research to continue, while at the same time limiting risky and inappropriate behavior and experiments.

As advances in synthetic biology continue, we must also be concerned with the potential for artificial organisms to be created. The implications of such a development cannot be overstated. Introducing a new "species" into nature could dramatically upset a tenuous biological balance. We have seen the potential for such havoc when natural, non-native species are introduced into an ecosystem. Examples include Burmese pythons and other large, invasive constrictor snakes in Florida and the Asian carp in waterways in the Chesapeake watershed. One study estimates that the total costs of invasive species in the United States amount to more than $100 billion each year.[21] Despite these disconcerting examples that suggest a potential future issue, artificially developed organisms are not now considered to be the most likely threat. In fact, a 2006 report based on the Lemon-Relman Committee concluded that the *de novo* synthesis of existing pathogens poses a greater near-term threat than the creation of artificial organisms through the parts-based approach."[22]

Still, other fields in biology remain largely in the early stages of understanding. Such is the case with the immune system. While many of the components and structures have been defined, the method of action for the immune system remains elusive. Over time, we have come to understand that a relationship exists between the messaging cytokines and the ability of the host to fight disease, but mechanisms of this regulation are not well understood. The cytokines serve an important function in signaling the immune system to defend against invaders identified as nonself, yet overstimulation of the immune system has the potential to cause significant damage to the host, creating an unchecked inflammatory response. For example, we have evidence that the reaction (or overreaction) of the immune system played a significant role in the 1918 Great Influenza, which disproportionally affected healthy young people whose immune systems worked feverishly to fight the infection. The overstimulation of the immune system caused more damage than the pathogenesis of the influenza virus. We are also learning that at times a more effective approach for treating infected people might well be to treat the response to the pathogen rather than to treat the pathogen directly. Likewise, as our knowledge of the immune system grows, we will also be able to field advanced diagnostics with the capacity to detect disease in a host in the presymptomatic or prodromal phase of the exposure. More-specific antibiotics will be made possible as both our knowledge of the immune system and bioinformatics couple to replace the largely trial-and-error system of drug development and replace it with a highly rational drug design methodology. Efforts will become more design driven rather than experiment driven. This means less trial and error. As we continue to expand our understanding of the immune system, we should expect to see greater "industrialization" of the fight against disease.

Greater knowledge of the immune system garnered through bioinformatics will also allow for a realization of the promise of personalized medicine. Instead of the very industrialized solutions whereby all receive the same treatments and doctors and patients alike hope for the best, we will be able to develop solutions that are compatible with the patient and highly targeted, providing a large therapeutic window. The adoption of systems biology approaches will accelerate these changes in the delivery of medicine with dramatic effects on quality and longevity of life.

RNA interference, more commonly shortened to RNAi, is another such field in which our knowledge continues to expand rapidly and which has benefited greatly by progress in bioinformatics. RNAi provides an innate defense against invading viruses. It also provides targeted, highly specific defenses against viruses that make it ideal for new-age therapies. RNAi works by silencing or preventing the expression of certain genes. This technology can be particularly useful in preventing the expression of harmful gene sequences for genetic disorders. However, the potential does exist for the misuse of RNAi. Consider if the genes associated with mounting an effective defense against a particular pathogen could be turned off such that the invading virus would be made more effective, or alternatively, if the efficacy of the host's immune system could be altered to turn off vital immune-system signaling and thus leave the host largely unprotected from disease. While we have learned much about RNAi, there is still much to learn before its full potential for use as a therapeutic can be realized. RNAi reflects yet one more technology that highlights the cautionary nature of the advances in biotechnology; that is, the same technology that has enormous potential for benefiting humankind also has the potential for catastrophic misuse, either through accidental or deliberate actions.

Progress in many areas of biotechnology is occurring rapidly and before all of the implications are fully understood. Given time and resources, this understanding will undoubtedly come. However, it will take several years until we will be able to predict with a high degree of confidence how pathogens interact with their hosts; this will require more knowledge in the field of proteomics. Therefore, for the foreseeable future, it will remain far easier to make slight manipulations of existing DNA than to synthesize an organism or virus. And while we have made great strides in the manipulation of traits in crops, these same processes are far less well developed in the manipulation of the traits of farm animals and even less well understood or developed for humans. Capabilities such as protein folding, modeling the immune systems, and understanding genetic interactions remain challenging today. However, over time, even complex tasks such as the development of an organism, as was recently accomplished at the J. Craig Venter Institute, will be reduced to industrial processes, but we are not there yet.

As our understanding of biotechnology increases, we are seeing an industrialization of the field that is leading to greater predictability. This has the potential for the types of benefits discussed above, yet also for those desiring to use disease as a weapon. One of the major criticisms from the military community in the pre-BWC days, at least from U.S. leaders, was the lack of predictability in the outcome of BW weapons. This reflected that BW weapons required an incubation period and were greatly impacted by environmental conditions, which meant that effects were not immediate but rather occurred over a period of time after an incubation and prodromal period that lasted days to weeks depending on pathogen, dose, and host immune system characteristics.

Today, many of these concepts may be outmoded in thinking about the potential for mortality and morbidity from a biological attack. For example, concepts such as LD50—lethal dose for 50 percent of the exposed population—are measures designed to characterize the virulence of a pathogen, yet they have little applicability for victims who might receive an overwhelming dose. Given such a dose (i.e., orders of magnitude higher than the LD50), the time for the effect of the attack to be seen may well be reduced well beyond historic time periods. For example, the LD50 for anthrax is approximately eight thousand bacteria; what if the number of bacteria provided to the host were eight hundred thousand bacteria, or two orders of magnitude greater than the LD50? Given such a dose, the number of infected cells functioning as bioreactors for the reproduction of *B. anthracis* bacteria would undoubtedly lead to the onset of symptoms well before the expected two-to-seven-days incubation period.

Concepts such as bioregulators were not considered at the time of the BWC EIF. Bioregulators are naturally occurring organic compounds that regulate diverse cellular processes associated with the nervous, endocrine, and immune systems. As such, they are related to substances normally found in the body that regulate these critical biological processes. Bioregulators have diverse chemical structures that range from hormones and neurotransmitters to complex macromodules such as proteins, polypeptides, and nucleic acids. The bioregulator industry has become big business, with literally tons of peptide bioregulators and their synthetic derivatives produced annually. One account lists applications including ailments such as "asthma, arthritis, cancer, diabetes, growth impairment, cardiovascular disease, inflammation, pain, epilepsy, gastrointestinal diseases, and obesity." This same account goes on to ominously warn, "Bioregulators could potentially be developed into biochemical weapons that incapacitate, alter moods, trigger psychological imbalances, cause a variety of other types of physiological reactions, or kill." Unlike traditional disease-causing BW agents that take hours or days to act, many bioregulators act within minutes of administration. Therefore, if exploited for the purpose of bioterrorism, they could potentially cause both profound and rapid physiologic effects, as their potential military use is similar to that of toxins that are fast acting and do not have an incubation period.

In the Industrial Age, chemical processes were used in the production of virtually everything from medicines to cars to fuel to computers. Today, biotechnology is quickly moving to replace many of these chemical processes, and it promises to do so at less cost and with less impact on the environment. Biofuels are rapidly becoming a reality. Biological products are replacing synthetic versions of heretofore chemically derived medicines. Even areas such as mining are seeing biological organisms incorporated with increasing frequency. Microbes are being used for environmental remediation as well, including cleaning up large oil spills using petrochemical scavenger bacteria.

To demonstrate the significant changes in the field of biotechnology, one need only look back at a Federation of American Scientists (FAS) study that examined progress in biotechnology from 1940 to 2000. The study listed fifteen components that had dual-use potential, including solid state peptide and nucleic acid synthesis, vaccines, nucleic acid probes, monoclonal anti-bodies, sensors, DNA engineering, pathogen efficacy, Human Genome Project, encapsulization and stabilization, antibiotics, bioactive peptides and cell growth chambers, and fermenters. The study asserted that changes in these key areas were occurring at a rate of 400 percent per annum by the year 2000. Not even included in the list are bioinformatics, DIY biology, BioBricks and iGEM, RNAi, bioregulators, genetically modified organisms, proteomics, metabolomics, nanotechnology, advanced manufacturing, and neuropharmacology. The difference in the lists reflects the changes that are occurring as biotechnology moves from a single, stove-piped field to a convergent activity composed of numerous complex, highly integrated fields.

As we close this section, we must conclude that the linkages and relationships between chemical and biological processes and hence between the BWC and the CWC are eroding at an alarming rate. We naturally find ourselves asking whether biopeptides and short-chain amino acids that are integral parts of the human bioregulator system should be characterized as bioarmaments or chemical armaments if used as weapons. The answer lies in the middle of an increasingly expanding gray area as these biological structures have chemical formulas, can be chemically derived, and can be synthetically developed. The answer to this question matters greatly as the CWC has an in-place verification regime, while the BWC does not. Still, even with this distinction, we must ask whether the CWC regime has the capacity to begin verification of biochemical material. To date, bioregulators have not been discussed within the CWC.

And finally, we must ask whether having separate regimes makes sense in the continuing convergence.

THE HUMAN DIMENSION OF THE BIOTECHNOLOGICAL REVOLUTION

Human history suggests that all major periods of technological innovation are accompanied by complementary periods of human adaptation. The Agrarian Age saw the development of modern city-states and the newfound importance of land ownership, while signaling a move away from the hunter-gatherer manner of human existence. The Industrial Age led to the increased productivity and industrialization of society but also a shift in the distribution toward urban and away from the agrarian life. The Information Age has introduced a new capacity for connecting people around the globe and sharing information; it has also introduced a new, largely ungoverned space we call the cyber domain. Today, we stand on the precipice at the Biotechnological Age where such fundamental change is possible that the very essence of life on our planet could be altered.

Yet as we approach this precipice, we found ourselves in a familiar position of not having the appropriate mechanisms to manage and control the dramatic changes that are unfolding in biotechnology. Enormous amounts of information are available that describe techniques and processes for manipulating biology, but these same documents do little to provide the necessary cautions about the danger of misuse. The industrialization of biology has placed greater capabilities at the hands of legitimate researchers and hackers alike, allowing for more experimentation and possibly more misuse. Soon, advanced techniques such as 3-D printing are likely to allow for unfettered (or at least virtually unlimited) access to dangerous pathogens.

Just as with the earlier periods, laws, regulations, and codes of conduct have not been written to ensure that biotechnological capabilities will be properly used. The 2011 H5N1 articles that caused a flurry of action within the U.S. government provides one data point concerning the Biotechnological Age. The two articles were based on National Institute of Health (NIH) sponsored studies that examined "gain of function" transmissibility in H5N1. Should the articles be allowed to be published, or was the information too dangerous to be made public? Were the experiments too dangerous to be conducted? Are there limits on the types and conduct of experiments? If so, who should set the standards, and on what should they be based?

Throughout the history of arms control, we have been conditioned to look at the control of things: material, equipment, knowledge, and systems. Export control regimes were designed to ensure that technologies and advanced capabilities were restricted and therefore not available to all. But what are the implications of new technologies such as the 3-D manufacturing discussed previously? How will these regimes need to be altered to protect against undesirable proliferation? Can this control realistically be established given that it is information that is now the coin of the realm? Using cyberspace, information including the genetic code for dangerous pathogens can be shared around the globe instantaneously along with the knowledge of how to process them into dangerous substances.

We must also ask ourselves if we are more concerned about the proliferation of high-containment laboratories (i.e., BSL-3 and -4) or the potential that experimentation with dangerous biological substances may be conducted without proper containment capabilities. In this debate, there is both a facilities question and one of experts policing themselves and their activities to assure that proper, safe science is being practiced. We must also be concerned about the proliferation of small, "ungoverned" community labs. This is not to say that small, privately owned labs are dangerous, but we must recognize that the same levels of protection, oversight, and codes of ethics do not normally exist in these facilities. Many of these DIY and biohackers have little or no formal training in risk education and certainly not risk mitigation. What dangers to the individuals and to society do these experiments pose?

To this point, one expert in biotechnology stated that he had become more concerned about bioerror than bioterror. In this quest for safe science, the requirements for international connectedness cannot be overstated. Dangerous or improper scientific techniques will likely have effects that will go well beyond the borders from which they originated. International efforts must ensure that basic standards of practice are established and adhered to across the globe. This means that international codes of conduct and practices must be developed and adhered to, early warning capabilities must be developed, and we must work to build partner capacity across the community.

In thinking toward the future, we can take an example from the development of software. Companies have teams of software developers around the globe, so that code can literally be developed twenty-four hours a day, with the results of a day's work passed to the next group of developers in a continuous cycle. If this same model were to be used by DIY biologists, we could find biological advances—some that would likely be quite dangerous, if misused—passed around the globe on a routine basis. Would it be possible to control the proliferation of such biological "material"? And what might be the effects of crowd sourcing for biotechnology as people look to gain greater information?

We must also be concerned with the competition for biotechnology that will undoubtedly occur for scarce biological resources. We have seen a glimpse of this future with the use of ethanol as an additive to gasoline. Ethanol is produced through an extraction process using corn as the basis for generating this source of energy. The competition comes down to whether the corn should be grown for human consumption or for use as a gasoline alternative. During normal climatic conditions, perhaps this does not present too great a challenge, but what about during periods of severe drought, such as during the summer of 2012 in the United States in which the shortages of corn translated into an increase in food prices?

Further, we must be cognizant that proliferation of biological capabilities is likely to result in industrial accidents with potentially significant potential. This was at the center of the H5N1 publication debate. This issue is not without precedent as the 1977 release of a H1N1 seasonal influenza strain and the 2003 to 2004 release of the SARS virus indicate. To deal with either the deliberate misuse or accidental release of biological material or the proliferation of potentially dangerous information, biorisk management techniques that rely on a combination of biosafety, biosecurity, and bioethics will become even more important.

We must also consider the workforce of the biotechnologist of the future. Today the field is composed primarily of molecular biologists. In the future and given the multidimensional aspects of biotechnology, chemistry, physics, mathematicians, statisticians, and information technology specialists are likely to feature more prominently in the field. As biotechnology becomes more industrialized, engineers will play just as pivotal a role as scientists. Subspecialties such as electrical and software engineering, microfluidics, data analytics, personalized medicine, and informatics will be essential to realizing the full potential of biotechnology.

Changes occurring in biotechnology are of great importance to our future and must be considered in light of required complementary changes to control regimes, laws, regulatory systems, and intellectual property mechanisms. While the science will allow for greater potential for advanced vaccines, drugs, and prophylaxis, the requirements of the Food and Drug Administration (FDA) do not lead to rapid licensure. Today, the only vaccine that receives rapid licensure is the seasonal influenza vaccine that gets approval in approximately sixty days, while other human vaccines require almost a ten-year period to gain approval.

In this new Age of Biotechnology, the importance of the intellectual property reform cannot be overstated. For the field to evolve within an open architecture, as has been the case with software development, patent reform, crowd sourcing, and sharing of information, to name a few, required changes will be necessary. However, this sharing of information, while having great potential for those conducting legitimate research, will serve as an additional proliferation window that will be difficult to control and virtually impossible to close.

We must deal with the question of pathogens and other biological material as intellectual property or IP. The original case that is being debated in U.S. courts is based on a lawsuit against Myriad Genetics in 2009 by the American Civil Liberties Union on behalf of twenty scientific organizations and patients. The specific issue was based on Myriad Genetic's request to patent two genes commonly associated with ovarian and breast cancers, the BRCA 1 and BRCA 2 mutations that increase breast cancer risk by 82 percent and ovarian cancer by 44 percent. This essentially gives Myriad Genetics control over all research and testing done nationwide on BRCA 1 and BRCA 2 mutations.

Arguments for and against being able to patent biological material such as genes are compelling. On the one hand, patents would limit the development and availability of tests, limit access to testing if a patent provider does not accept a patient's insurance, and limit multigene testing, which has important potential future benefits. The contrary argument is that failure to allow companies to patent genes reduces the economic incentive to engage in research. One account provides the following narrative:

> Civil libertarians and patent opponents object to companies claiming they have invented what nature has wrought and contend that such patents hinder life-saving research. Corporate patent owners say their scientific ingenuity is

needed to isolate the genes and that research could stall without the protection of patents.[23]

In looking to the future of governance measures to be considered, one framework presents a three-pronged model that includes hard law, soft law, and informal mechanisms. The hard law consists of civil and criminal statutes, statutory regulations, export controls, reporting requirements, and mandatory licensing, certification, and registration; soft law consists of security guidelines, industry or scientific self-governance mechanisms, adoption of international standards and prepublication review; and informal mechanisms consist of codes of conduct, risk education and awareness training, whistle-blowing channels, and transparency measures.

In this spectrum of governance, the measures that are most closely related to the scientists are the ones more likely to be successful. For example, proper training and implementation of codes of ethics are far more likely to be successful in limiting dangerous proliferation than export control and laws. For this reason, documents such as the *Biosafety in Microbiological and Biomedical Laboratories (BMBL)*, first published in 1984, and the National Science Advisory Board for Biosecurity (NSABB) established after the 2004 Fink Report on responsible life sciences research are critical to biorisk management (which includes biosafety, biosecurity, and bioethics) practice and policy in the United States.

This "hard law, soft law, informal mechanisms" approach seems to make a lot of sense, but we must ask, "Is it too rigid and linear given the very nonlinear and rapid developments in biotechnology and its convergent technologies?" Consider the IP case discussed above. It has been ongoing since 2009 and as of the publication of this text in 2013, it had not reached a conclusion. Meanwhile, biotechnology has continued to advance at an exponential pace.

THE ROAD TO THE BWC

In 1969, President Nixon renounced the offensive use of biological weapons. It was also the first time that an enzyme was synthesized *in vitro* for the first time. The state of biotechnology was in its infancy, with little known about the inner workings of the basic forms of life. Recombinant technology was more concept than practice, and it would be several years until a successful recombinant experiment would be conducted.

The U.S. program had revealed the potential of biological weapons to be used for offensive purposes and suggested that biological weapons could have strategic impact if properly prepared and used against an appropriate target. Yet the U.S offensive program also revealed that effective BW programs required a great deal of care in order to prepare these biological pathogens for release and appropriate dissemination. Specialists with very unique training in areas such as preparation of media for growing pathogenic living organisms, drying the biological preparations while keeping the agent viable, milling the agent to the appropriate size, ordnance development, and disseminating the product without destroying the biological material required great care.

The offensive programs, declared and undeclared, of the major powers provided an instructive road map that served to highlight the major issues of concern at the time. The list of pathogens that were seriously considered for use included some ten or so agents with very familiar names. In the Soviet program, these agents included smallpox, plague, anthrax, botulinum toxins, Venezuelan equine encephalitis (VEE) virus, tularemia, Q fever, Marburg, influenza, melioidosis, and typhus. The United States list was similar yet contained some notable differences, including the incorporation of plant pathogens such as wheat stem rust, rye stem rust, rice blast, and the human pathogens brucellosis and *Staphylococcus enterotoxin* B (SEB).

Conclusions from the former offensive programs led to deductions based on biotechnology at the time and not necessarily the potential for an offensive BW program in the twenty-first century. These same agents became known as especially dangerous pathogens and received extra scrutiny. The thought at the time was that biological weapons were strategic weapons and had significant drawbacks for use as battlefield weapons. The incubation period was thought to make these types of weapons not useful for achieving a battlefield effect. BW experts, but not necessarily policymakers or even those in the medical profession who were not expert in infectious disease, understood concepts such as overwhelming dose.

At the time of entry into force of the BWC, the first recombinant DNA experiments had just been concluded using bacterial genes; polymerase chain reaction (PCR) amplification of DNA was almost a decade away. Proteomics (the branch of biochemistry concerned with the structure and analysis of the proteins found in living organisms) and metabolomics (which is the study of all the metabolites present in cells, tissues, and organs) had not been conceived. Synthetic biology, which is the "product of two different convergences: between biology and chemistry and between engineering and living systems," had not been developed.[24]

Given the state of biotechnology at the time of BWC EIF and the influence of the offensive BW programs, it should come as no surprise that the Convention's focus would be on state biological warfare programs involving especially dangerous pathogens that had the capacity to inflict high mortality and morbidity in a strategic attack on unsuspecting populations.

Equally predictable is the derivation of the confidence-building measures (CBMs) for the BWC. Given the envisioned threats at the time of the development of the CBMs and the focus on state BW programs, these measures were designed to focus on high containment facilities that states would likely use in the development of an offensive BW program.

At the same time, the state of biotechnology has so fundamentally changed that the utility of the CBMs must be carefully considered to ensure viability well into the future. The threat comes from sources and materials that were not even contemplated at the time of the BWC EIF. This is not to say that the BWC does not have relevance, but rather that the center of gravity or focus must shift to new areas of concern. Additionally, the systemization of the biotechnology field—the use of engineering techniques to make routine the biotechnological processes—has converted biotechnology into a science rather than an art. Whereas previously, scientists had to understand such techniques such as polymerase chain reaction (PCR), today, one needs only understand the concept so that the output can be generated. A machine is capable of amplification of the DNA whether or not the scientist understands the scientific steps required of the process.

DEMONSTRATING THE COMPLEXITY OF THE ISSUE

Scenarios can be useful to illustrate the inherent complexity with understanding the relationship between biotechnology and the BWC. For this analysis, we will examine a range of uses of biological capabilities and process and then assess whether or not these uses constitute a violation of Article I of the Convention.

In the first case, a nation uses petrochemical-eating bacteria for remediation of an oil spill. The use is clearly justified for "prophylactic, protective or other peaceful purposes," and therefore it is permitted under the BWC. What if these same bacteria were deployed along a nation's border in a defensive line where they could destroy the tires of an advancing enemy? Could the nation deploying the bacteria make the case that the use of these bacteria was to protect against the attackers breaching its national borders and therefore is permitted under the BWC? At the very least, this would have to be considered a gray area and subject to interpretation and debate. Now let's take the scenario one step further and say that a nation that was about to be attacked decided as a protective measure to covertly deploy the bacteria in the territory of the anticipated attacker? While the intent may have been protective, the preemptive attack would strongly suggest a violation of the BWC.

In another case, consider a biohacker with no particular intent other than to learn more about biology who inadvertently develops a mutation of H5N1 avian influenza that increases the virulence of the pathogen, giving it higher invasiveness and toxigenicity. The derived pathogen escapes from the poorly constructed facility and infects thousands, causing ten thousand hospitalizations and leading to 60 percent mortality or six thousand deaths for those who are hospitalized. The state where the biohacker lived was a member of the BWC, yet the state did not have any laws governing working with highly pathogenic biological material. Has there been a violation of the BWC by the biohacker? By the state? In answering this question, we must remember that the BWC is a treaty signed by states governing the actions of other states. What degree of culpability does this state have as a result of the subsequent human, societal, and economic toll? Would that culpability be less if the state had proper laws on the books and national enforcement mechanisms?

For our next case, consider the implications of a state that uses a particular transgenic variant of wheat, knowing that its use will likely adversely affect the wheat crops and future yields of a neighboring country. As predicted by the science, the country that uses the modified wheat experiences record yields, while the neighboring nation sees its crop yield drop dramatically. Does this constitute a violation of the BWC? Can the neighboring nation declare a biological attack, or is this the result of extreme economic competition? It would be difficult to demonstrate that the use of the transgenic wheat was intended to decimate the wheat yields of its neighbor; however, prudence would dictate that caution be used prior to use of a GMO.

THE REALITIES OF THE BIOENVIRONMENT

If we juxtapose the likely biological threats with the greatest opportunities for the BWC to contribute to global security, one gets an interesting view of the opportunities and challenges we are likely to face in the future bioenvironment.

Examining the historical perspective, the dual-use concerns and the likelihood of certain types of events provides for interesting conclusions related to biotechnology and the BWC. The history of the BWC period indicates the focus on state BW programs in the period immediately prior to, during, and immediately after the BWC EIF. More recently, we have seen the use, or attempted use, of BW by bioterrorists, albeit not with particularly impressive results. The limiting factor in several of the high-profile cases has been the availability of the pathogen or seed stock for use in a biological weapon.

With increased globalization, we continue to see the emergence of new naturally occurring disease and global pandemics. This should come as little surprise given the harbingers of such activities seen through global patterns of movement and disease going back to the plague in the Middle Ages, the waves of smallpox pandemics, and the 1918 Great Influenza. Today, we are seeing more diseases transiting the globe with increased regularity and rapidity. The seam between the human and animal kingdoms is also a place of great concern, for where there is close contact, we have seen the potential for devastating consequences including with diseases such as Ebola and SARS.

Consider West Nile Virus, which had not been seen in the United States until the late 1990s and now is endemic in many regions of the country. As human populations continue to mix with even more frequency and humans continue to encroach upon naïve habitats, we should expect that these trends continue at an even more accelerated pace.

We should also expect that the proliferation of biotechnology will allow for more advanced technical means, including material and information, to be available to a greater percentage of the population. As we have seen, albeit not on a large scale to date, there is interest in BW capabilities by terrorists or lone-wolf actors. The overarching trend will be the proliferation of statelike capabilities in the hands of nonstate actors.

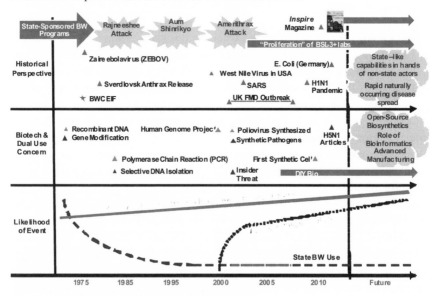

Figure 2.1.   The Nature of the Biological Threat

Examining the pace of biotechnology, we are seeing exponential growth in capabilities combined with a convergence of several fields, including engineering, mathematics, and information technology with biotechnology. In the classic dual-use sense, we are going to see more opportunities to conduct experiments and develop capabilities that were previously only possible by experts. Areas that were only a few years ago considered to be worthy of a Nobel Prize are being incorporated into junior high and high school classrooms. More information is available online, and there are few tangible controls on the proliferation of information. Biotechnology is moving from a scientific to an engineering discipline for many heretofore complex tasks. While the cutting edge of biotechnology remains a scientific discipline, much of the field is being reduced to the solving of engineering problems. All of this is to indicate that the ability to develop a BW capability will move from being an art, understood only by experts such as William Patrick from the former U.S. offensive BW program, to a replicable engineering outcome.

In looking at the range of biological threats in three categories—the state use of BW; an accident, misuse, or bioterror; or naturally occuring biological event—the most likely is the naturally occurring event followed by the potential for an accident, misuse, or bioterror. State use is considered a distant third.

As we consider the relevance of the BWC, we should ask, how can and how should the BWC play a role in this new era?

CONCLUSIONS

Biotechnology today is in a period whereby it is transitioning from a scientific to an engineering discipline. This has important implications for even greater discovery and benefits for humankind. Yet it also provides a lowering of thresholds for those (states and individuals) attempting to use biotechnology for nefarious purposes as well as for the potential inadvertent misuse of biotechnology. The dual-use conundrum is coming more clearly into focus as this evolution continues.

For the last twenty years, many of the advances in biotechnology have been linked to the development of the polymerase chain reaction process used for amplification of DNA. PCR has been used in a wide variety of essential biological endeavors and enabling technologies such as "DNA cloning for sequencing, DNA-based phylogeny, or functional analysis of genes; the diagnosis of hereditary diseases; the identification of genetic fingerprints (used in forensic sciences and paternity testing); and the detection and diagnosis of infectious diseases" to name a few.[25] Perhaps the more relevant question for us is what the next PCR-like game-changing technology is and when we will see it incorporated into biotechnology.

At this point, a note of caution is in order. This chapter has been about advances in the field of biotechnology. It was written to look at the field in a futuristic sense, to look at the possibilities that are unfolding before us. Where applicable, the potential for misuse of biology was considered to at least raise the possibility of an attack using advanced biotechnologically developed capabilities. However, this is not meant to imply that this is the most likely scenario for a pathogen used in an attack. Perhaps it has become cliché, but it is worth repeating that Mother Nature remains the best bioterrorist. Pathogens found in nature have evolved to provide the capacity to cause illness and death and require little, if any, modification or augmentation. Consider that the combination of HIV/AIDS, malaria, and tuberculosis that causes over six million deaths annually around the globe. Naturally occurring diseases such as foot-and-mouth disease (FMD) in livestock have the potential to cripple economies and threaten our food supply in their natural state and require no modification.

While the participants at various BWC events, including the review conferences and intersessional meetings, have continued to emphasize that progress in science and technology has not invalidated the basic articles of the convention, we must be concerned about the BWC having the structures, organizations, and processes to adequately monitor, track, and address the changes that are unfolding in this emerging Age of Biotechnology. The genius of the BWC was the broad and enduring nature of the convention that provided a vehicle for controlling dangerous state biological activities yet not stifling scientific discovery.

The BWC has evolved from the early 1970s, from the knowledge of the biotechnology field at the time it was negotiated and largely based on the potential for biological pathogens as understood based on state offensive programs at the time. During the period from BWC EIF to the present, we have seen biotechnology advance at rates that one study estimated were 400 percent per year.[26]

Over time, the basic Articles of the Convention have been augmented with data exchange requirements through voluntary confidence-building submissions, yet little else has been done to change or update the treaty. As the transition of biotechnology from science to engineering continues, examination of the implications for the BWC is indeed in order.

Chapter Three

Ten Reasons Why the BWC Matters

The premise of this book is that the BWC is the most important arms control treaty of the twenty-first century. The convention operates at the very intersection of national security, arms control, and biotechnology, and globalization serves as the catalyst for these interactions.

It might be tempting to think of the BWC as an apparatus of national security or a simple arms control treaty. Yet that would be "missing the forest for the trees," as the saying goes. There is so much more to the BWC that needs to be examined and understood in light of changes in our biotechnology landscape. This chapter will do just that.

Globalization has made the world smaller with a greater number of human interactions. Peoples and cultures routinely interact with each other and new environments. Some of these interactions lead inevitably to tensions and potentially conflicts, as well as exposing people to a broader range of threats. Competition for resources, conflicts fueled by the clash of cultures, and the opening of areas that heretofore have been isolated are a direct result of globalization. Information streaming unfettered across the globe and being shared in real-time creates a sense of connectedness and disconnectedness. There are winners and losers, and the results play out instantaneously. It really does matter what happens across the street and across the globe.

The sheer magnitude of the biotechnology industry mandates that it will be of growing importance to our livelihood, including our health, welfare, security, and ultimately our longevity. Biotechnology shows great promise, but as the history of humankind demonstrates, virtually no technology is immune from misuse or use for nefarious purposes.

The effects of the BWC are pervasive. The degree to which the Convention touches the lives of average citizens, as well as the national leaders responsible for homeland and national security, should not be underestimated. When you go about your daily business, you assume a certain degree of protection, certainly free from attack by a state or a nonstate actor. Nor do you expect to be attacked by biological weapons. You understand that naturally occurring disease such as the common cold is somewhat inevitable, but you expect that governments will step in to protect populations and to find solutions related to a large-scale biological event across the full range of anticipated threats.

The Presidential Policy Directive (PPD)-8, National Preparedness, aims to "strengthen the security and resilience" of the United States through "systematic preparation for the threats that pose the greatest risk to the security of the Nation." At the operational level the PPD provides a framework for security planners, governments at all levels, and first responders to develop the capabilities necessary to prevent, protect, mitigate, respond, and recover from major threats; this includes against a bioevent, either naturally occurring or man-made.[1] While the phrase *Biological and Toxin Weapons Convention, BWC*, or any other related term does not appear within PPD-8, clearly a biological event has the potential to fall within the PPD-8 realm.

So with that as an introduction, here are ten reasons why the Convention matters. The Biological and Toxin Weapons Convention (BWC) . . .

1. ...eliminates an entire class of dangerous weapons.

The BWC often gets credit for being the first arms control to eliminate an entire class of weapons. This bumper sticker is both true and also a little misleading, as the Convention does not prohibit research on biological agents so long as the work is being pursued for "prophylactic, protective or other peaceful purposes" such as the development of medical countermeasures and health protection for our military. But the focus of this assertion is less on the elimination of a class of weapons than on the concern about the danger posed by biological weapons; that is, the agent plus the delivery system.

On this, considerable debate has ensued. Some believe that the threat posed by biological weapons has been hyped. Others believe that this threat is one of the most dangerous and challenging that we face today as a nation and, indeed, globally. The history of the former state programs strongly suggests that the latter is correct and that developing preparedness and response capabilities is imperative. In 2009, the White House released the *National Strategy for Countering Biological Threats*, which characterized the threat of a biological attack as placing at risk the lives of hundreds of thousands of people.

Scenarios abound in trying to articulate the threat posed by biological weapons. A 1970 World Health Organization (WHO) expert estimated casualties following a theoretical release of fifty kilograms of anthrax spores from an aircraft over an urban population of five million people. It estimates there would be 250,000 casualties, of which 100,000 would die without proper treatment.[2] A later 1993 report by the U.S. Congressional Office of Technology Assessment looked at a scenario involving the release of one hundred kilograms of anthrax aerosol upwind of the Washington, D.C., area. It estimated that this would cause at least 130,000 deaths and possibly as many as three million. An economic model developed by the Centers for Disease Control and Prevention (CDC) estimated a cost of $26.2 billion per one hundred thousand people exposed to a bioterrorist attack.[3]

The late William Patrick, who was chief of product development in the U.S. offensive BW program until the 1969 decision to renounce these weapons, believed that BW was a significant and very misunderstood threat. Shortly before his death, Patrick encapsulated his thoughts on this form of warfare, making five claims. First, he believed that BW had the capacity for delivering nuclear equivalence in the same way that the neutron bomb was postulated to be able to kill large populations without destroying the cities. Second, he articulated the concept that not all agents were considered equal; with proper processing and weaponization, highly lethal biological pathogens could be made even more lethal. Third, he stated that pathogens such as *Francisella tularensis* and *Staphylococcal enterotoxin* B (SEB) deployed in combination could magnify the effects and create what he came to call "killing winds." Fourth, he stated that overwhelming doses of BW acted not as naturally occurring infectious disease but rather as weapons in which historic incubation times and mortality/morbidity rates would not apply. Finally, he believed that deterrence in a classic sense was not possible because of the uncertainty of attribution.[4]

Other tests done as part of assessments to determine the threat posed by BW provide ominous warnings. The Department of Defense used special operations forces from Fort Detrick to run "vulnerability tests" using suitcases to disperse *Serratia marcescens* and a later experiment in 1966 using three *Bacillus subtilis* filled light bulbs tossed into a New York subway to gauge the effects and dispersion patterns. In the case of the latter experiment, the assessment was that had actual anthrax been used rather that the simulant, a million passengers would have been killed.[5]

The belief, expressed by then-Speaker of the Iranian Parliament Hojjat al-Islam Akbar Hashemi Rafsanjani in 1995, that BW is "the poor man's atomic bomb"[6] demonstrates a pattern of thinking about the potential for biological weapons. The low cost, ready availability of naturally occurring pathogenic seed stock and BW's highly destructive potential have contributed to this mantra. Perhaps this is a bit of hyperbole, but it does give one cause for concern.

One project conducted in 1999 to 2000 by the Defense Threat Reduction Agency (DTRA) was designed to assess the potential for a terrorist to develop a BW weapon. The project, code-named Project Bacchus, highlighted the ease of development of a small-scale BW capability: [7]

> Operating out of a former barber shop and recreation hall without arousing any suspicion, the team purchased the necessary glassware, piping and filters from a local hardware store. It also ordered a 50-quart fermentation unit from Europe for growing the bacteria, high-ranking Pentagon officials said yesterday. The officials said the team purchased a milling machine capable of grinding dried material into fine powder from a store in the Midwest. The team was not allowed to produce real stains of anthrax, but only biopesticides during two production tests in 1999 and 2000. Pentagon officials who studied the results of the test said the scientists, with anthrax spores, could have produced enough of the bacteria to have killed at least 10,000 people. The scientists succeeded in developing a lab capable of producing bacteria that could kill thousands of people, and did so on a budget of about $1.5 million.

The Obama administration in 2009 issued the *National Strategy for Countering Biological Threats* and an accompanying Presidential Policy Directive (PPD)-2 for its implementation. [8] In a briefing done by the National Security Council staff, a contrast was drawn between a small-scale attack, such as the 2001 anthrax attacks, with a postulated medium-scale aerosol release. The 2001 anthrax attacks, commonly referred to as the Amerithrax attacks, required thirty thousand people to receive antibiotic prophylaxis, twenty-two were made ill, five died, and six buildings needed to be decontaminated at a cost of $1 billion. In the postulated medium-scale release using Washington, D.C., as the target, the estimates were that 1.9 to 3.4 million people would require antibiotic treatment, approximately 450,000 would be made ill, and 380,000 of those would die, and the city would need to be decontaminated at a cost of $1.8 trillion. [9]

The descriptions of the Soviet and then the Russian program following the end of the Cold War are even more chilling, with descriptions of large stockpiles, extensive open-air testing, and a doctrine for using BW weapons against populations to achieve their strategic objectives. Much of what we first learned about the Soviet BW program came from defectors such as Vladimir Artemovich Pasechnik, who was a senior Soviet biologist; bioweaponeer Ken Alibek, who was the deputy for Biopreparat; and Sergei Popov, who served as a division head at the State Research Center of Virology and Biotechnology (known as VECTOR) and at Obolensk, both branches of the Soviet bioweapons program dedicated to developing genetically enhanced products. These depictions were later confirmed, largely through our Nunn-Lugar Cooperative Threat Reduction (CTR) efforts.

One account of the Soviet program describes a sixty-thousand-person effort under the leadership of Biopreparat, which included large-scale production facilities and massive stocks of biological pathogens, including

"1,500 metric tons of tularemia bacteria; 4,500 metric tons of anthrax; 1,500 metric tons of bubonic plague; and 2,000 tons of glanders bacteria."[10] One report on the Soviet program suggested that they actually made such biological modifications as part of their state BW program; in one such experiment, they were known to have genetically engineered a combination of *Variola major* (smallpox) and Venezuelan equine encephalitis (VEE).[11] Regardless of the exact totals, suffice it to say that the program was extensive.

No less chilling, and certainly an indication of the regard with which the Soviets held BW, was the statement by Soviet Marshall Georgi Zhukov. In 1956, Zhukov gave a speech, in which he stated that "future war, if they unleash it, will be characterized by . . . atomic, thermonuclear, chemical and bacteriological weapons."[12] Following Zhukov's lead, the Soviets had developed what has been characterized by one author as "The Dead Hand," which details the degree to which the Soviet Union had planned to incorporate the full range of weapons of mass destruction (WMD) into their strategic, operational, and tactical planning.[13]

While reports have periodically surfaced about BWC noncompliance, there has been no confirmed state use of BW in an offensive manner since the BWC entered into effect. Accidents such as the 1979 Soviet inadvertent release of anthrax at the Sverdlovsk production facility indicate that not all have complied with the articles of the BWC. Therefore, on balance, we can conclude that the BWC contributions to the elimination of an entire class of dangerous weapons have been important.

2. ...provides an unequivocal norm against the use of biological weapons.

Perhaps even more important than the elimination of a class of dangerous weapons has been the unequivocal norm that has been established concerning the use of biological weapons. Early thinkers have recognized this repulsion to BW. Vannevar Bush described it in 1949,

> Without a shadow of a doubt there is something in man's make-up that causes him to hesitate when at the point of bringing war to his enemy by poisoning him or his cattle and crops or spreading disease . . . The human race shrinks and draws back when the subject is broached. It always has, and it probably always will.[14]

Many believe this norm against BW production has proven more effective than other verifiable arms control measures.[15] However, others have argued that the normative taboos are "inadequate and potentially dangerous," as it has become the *de facto* basis for the BW nonproliferation regime and has forestalled efforts to compose more substantive arms control measures with effective verification mechanisms.[16]

The history of BW accusations and use can be insightful in this regard. On one hand history tells us that the number of nations engaging in offensive programs, at least at the research and development (R&D) stage, has actually increased since the BWC entered into force. At the time of the BWC entering into force in 1975, four nations were believed to have BW, including the United States, Soviet Union, China, and South Africa.[17] In 1989, Central Intelligence Agency Director William Webster announced that "at least 10 countries" were developing BW weapons.[18] In 2005, one source suggests that seven nations currently have programs, including China, Egypt, Iran, Israel, North Korea, Russia, and Syria.[19] Another indicated concern about "compliance for ten or so countries with the treaty and concerns about the biological weapons programs in a few countries that are not party to the treaty."[20] A more recent 2011 State Department unclassified report on disarmament treaty compliance stated that North Korea may consider BW as an option and expressed uncertainty about whether Syria and Iran were conducting activities prohibited by BWC; this same report found no evidence of noncompliance for China and Russia. The good news is that the number of suspect nations has gone down; however, this decrease does not necessarily imply greater compliance, as it could also be related to limited reporting and lack of information.

In examining this question of state proliferation in 1995, the Office of Technology Assessment at Senate hearings listed seventeen countries that were "suspected of manufacturing biological weapons."[21] This statement reflects an increase in the estimated number of likely BW-capable states at the time of the signing of the BWC and the statement by CIA Director Webster in 1989. The question of what was the actual number of BW-capable states was then and remains now a difficult one, as the data indicates. The lack of a viable BWC verification regime coupled with the ease of masking a BW program has contributed to this shortfall.

Turning to the question of the potential for a terrorist BW event, one author describes a risk analysis framework for examining the potential for terrorist use of BW. Jessica Stern, a noted terrorism expert, developed several key themes that have great applicability for understanding the motivations and possible moderating behaviors that might influence a terrorist. One can imagine the loss of confidence in the government should a successful, large-scale BW attack sicken and kill thousands of citizens. Could the government survive such a crisis in confidence? What would be the response? Stern also introduces the concept of "dreaded risks," a category of risk in which the fear of an act is disproportionate to the actual outcome, and she identifies BW in this category as a means that causes fear, angst, and reactions disproportionate to the actual potential of these capabilities. Continuing with this line of reasoning, Stern identifies four aspects of dread, which she labels as disgust, horror of disease, loss of faith in the ability of scientists to protect us, and implications for risk analysis and policy. Disgust stems from the idea of involuntary exposure. Horror of the disease is based on the inherent fear of disease and possible contagion. The loss of faith in the ability of scientists to protect pertains to both the idea of inevitability and the relatively new concept that science can be used for nefarious purposes, which diminishes our "technological optimism."[22]

In examining the BWC as a norm, we see a mixed result. We have already identified that the BWC did little to stop several nations from pursuing some type of BW program. In most cases, the nature of the suspected programs is not well understood. They could be offensive research and development programs. They could be larger-scale programs that have progressed to stockpiling and preparation for potential use. They could also be legitimate biodefense programs that fall within the acceptable range of the BWC that allows biodefense activities for "prophylactic, protective or other peaceful purposes."

However, we must also acknowledge that we have not seen widespread use of BW. Since the 1940s, a National Academy of Science (NAS) report lists seven instances of the use of BW weapons, including the 1940 to 1941 Japanese use of BW in China in the Hangzhou and Nanjing provinces; 1957 to 1963 in Brazil against Indian tribal populations; 1981 in the United Kingdom when commandos used anthrax against a research facility; 1984 in Oregon, United States, when the Rajneeshee cultists tried several BW attacks; 1989 in Namibia during a covert operation by South Africa; 1990 to 1993 in Japan by the Aum Shinrikyo cult using a variety of agents; and 2001 in the United States using anthrax-filled letters to attack a number of targets.[23] Of these seven instances, five occurred after the BWC entry into force, two were state uses of BW, and three were terrorist attacks. Of course, this listing of cases does not include the alleged assassinations or the thousands of "white powder" hoaxes that have occurred.

Arguably, the two most robust BW programs that have been identified in the post-BWC EIF era are the Soviet (that extended into the Russian era) and the Iraqi programs; notably both nations were party to the BWC during the time of their offensive programs. The case of the Soviet program has already been presented and will not be revisited except to say that it dwarfed all other state programs by orders of magnitude. The Iraqi program was unique in that Saddam Hussein's forces actually had deployed BW weapons in their arsenal and prepared for their use at the time of the first Gulf War in 1990 to 1991. "Arguably" is used as a qualifier, as little is known definitively about programs in North Korea, Syria, and Iran.

Still, when one considers the two small-scale uses of BW identified in the NAS report and the lack of evidence of state use, one must conclude that BW has not been seen as an effective instrument of warfare by a state. Attributing this solely to the BWC would be an overstatement, but certainly the BWC has had an influence on state behavior.

With respect to terrorism, the "norm" has not been nearly so effective. The BWC is a state-to-state, international arms control treaty with no direct link to terrorist behavior. Additionally, the nature of terrorism is to use a variety of means in order to promote causes and gain recognition. This has included the pursuit and in several cases the use of BW. To date, the attempts have been small scale, not well perpetrated and not particularly successful in terms of mortality and morbidity. In one alarming call to arms, al-Qaeda radical cleric Anwar al-Awlaki stated in an article that "the killing of women and children and the use of chemical and biological weapons in addition to bombings and gun attacks"[24] is appropriate and even encouraged. In an earlier version of *Inspire*, an online al-Qaeda magazine, the authors called for chemists and microbiologists to develop weapons and to attack the West. Still, al-Qaeda's biological program looks to be more aspirational rather than a well-established developmental effort.

For example, in the aftermath of the Mumbai terrorist attack, it was reported that:

> [T]he government had expressed concern about India becoming a target of a biological terror attack, with fatal diseases such as anthrax being released into the country before spreading around the world. A senior Indian diplomat told the George W. Bush Administration in 2006 that concerns about biological weapons were "no longer academic", adding that intelligence suggested terror groups were increasingly discussing bio-warfare.

The reference goes on to state that then Secretary for International Organizations, K.C. Singh further elaborated that "Indian intelligence was picking up chatter indicating jihadi groups were interested in bio-terrorism and seeking out 'like-minded PhDs (researchers) in biology and biotechnology'." The reference to seeking like-minded PhDs refers to the original *Inspire* magazine article that calls for the use of chemical and biological weapons.

So while the norm against the use of BW appears to have had some moderating effect on states, the same cannot be said for terrorists for which aspirations for gaining a BW capability are being openly considered. These both suggest an issue with the BWC and a potential solution required for strengthening the norm against nonstate actors.

3. ...provides an important international venue concerning biological defense issues.

U.S. national security policy is based on a layered approach that includes international laws and treaties, international organizations, voluntary nonproliferation consortiums that are essentially coalitions of the willing, and U.S. unilateral actions and programs. These bins of activity are further organized along functional lines whereby communities of interest form that focus on particular issues. Such is the case with *biodefense*, a broad term which has come to include a wide range of activities under an umbrella term of *One Health*.

At the lowest levels, national laws, policies, regulations, and programs in public health, medicine, biotechnology, and for handling highly pathogenic material or Select Agents (SA) serve to limit behavior and provide direction on biological issues. While these national arrangements serve a vital role in providing for the health and welfare of Americans, they can also be directly traced to international forums that provide the strategic underpinnings on which our national biodefense program rests. This includes the BWC.

Coalitions of the willing in biodefense include such entities as the Australia Group. The Australia Group is "an informal forum of countries" designed to limit the export of materials that could potentially aid in the development of chemical or biological weapons. Over forty nations are participants in the Australia Group, and nations are members of both the CWC and BWC. The group limits through licensing measures the proliferation of certain precursors of chemical, biological agents, dual-use equipment, and capabilities. In this regard, the Australia Group works in concert with the BWC and is seen to have direct linkages to several of the articles of the Convention.[25] Another example of a coalition of the willing is the Proliferation Security Initiative (PSI) that aims to stop trafficking of weapons of mass destruction (WMD), their delivery systems and related materials to and from states, and nonstate actors of proliferation concern.[26] Over ninety nations have endorsed PSI and the accompanying interdiction principles that call on participants to: "interdict transfers to and from states and non-state actors of proliferation concern to the extent of their capabilities and legal authorities; develop procedures to facilitate exchange of information with other countries; strengthen national legal authorities to facilitate interdiction; and take specific actions in support of interdiction efforts."[27] As with the Australia Group, the PSI traces its roots, at least in part, to the same nonproliferation outcomes as the BWC.

International organizations such as the World Health Organization (WHO), Food and Agriculture Organization of the United Nations (FAO), and the World Organization for Animal Health (OIE) have a direct and growing role with the BWC. These expanding relationships have resulted from the recognition that strong linkages exist between biodefense and efforts to promote human, animal, and plant health for the overall benefit of humanity.

The WHO directs and coordinates authority for health within the United Nations system and is responsible for providing leadership on global health matters, shaping the health research agenda, setting norms and standards, articulating evidence-based policy options, providing technical support to countries, and monitoring and assessing health trends. In the twenty-first century, health has become a shared responsibility, involving equitable access to essential care and collective defense against transnational threats.[28] The WHO also has developed and promoted strong epidemiological capacity for identifying, tracking, and disseminating information on emerging infectious disease. It is in this role that we see an area of significant overlap with the BWC.

The FAO supports the international community by providing access to information, sharing policy expertise, providing a meeting place for nations, and bringing knowledge to the field concerning issues related to food supplies. Topics of interest to the FAO include issues such as rural development, avian influenza, biodiversity, food safety, bioenergy, and food security, to name a few.[29]

The OIE lists six missions as part of its charter: ensure transparency in the global animal disease situation; collect, analyze, and disseminate veterinary scientific information; encourage international solidarity in the control of animal diseases; safeguard world trade by publishing health standards for international trade in animals and animal products; improve the legal framework and resources of national veterinary services; and to provide a better guarantee of food of animal origin and to promote animal welfare through a science-based approach.[30] Topical areas of interest are also of concern to the OIE, such as invasive alien animal species, emerging infectious disease such as the Schmallenberg virus, and bioterrorism.

One need only compare the national obligations of participating in the OIE with those of the BWC to see the strong overlap. For example, the OIE requires that "each Member Country undertakes to report the animal diseases that it detects on its territory."[31] In a related and complementary requirement, the BWC, through the Confidence Building Measures (CBM) annual submission under Measure B, requires, "Exchange of information on outbreaks of infectious diseases and similar occurrences caused by toxins that seem to deviate from the normal pattern." The requirements for reporting are very similar; the reasons for requesting such a report vary greatly. In the case of the OIE, the desire is to protect the global animal health, while for the BWC, the focus is to identify if a biological attack has occurred.

This increasing collaboration among the WHO, FAO, and OIE, as well as among these three organizations and the BWC can be seen clearly in a tripartite note between the WHO, FAO, and OIE that states,

> Pathogens circulating in animal populations can threaten both animal and human health, and thus both the animal and human health sectors have a stake in, and responsibility for, their control. Pathogens—viruses, bacteria or parasites—have evolved and perfected their life cycles in an environment that is

more and more favorable to them and ensures their continuity through time by replicating and moving from diseased host to a susceptible new host.[32]

With the growing relationships between the WHO, FAO, and OIE and the BWC, some have expressed concern, calling the new relationships a step toward the securitization of health. On the other hand, these growing relationships are a natural outgrowth of the globalization of society where the distinction between a naturally occurring outbreak and a deliberate biological attack may be difficult and in some cases nearly impossible to discern. Given the growing importance of these international collaborations, it will come as little surprise that representatives of the three health-specific organizations have begun participating in BWC-related activities, including in the seventh BWC Review Conference in December 2011.

There is good reason to bring the human, animal, and plant biodefense communities into greater alignment and synchronization of activities. From 1940 to 2004, 335 new emerging infectious diseases (EIDs) were identified. Of those, 60 percent are zoonotic, and many of these are some of the most deadly pathogens known, including Ebola and Marburg and even more recently the Nipah and Hendra viruses.

In yet another important example of the intersection between the WHO and BWC with positive outcomes, one need only consider the Internal Health Regulations (IHR). The WHO's IHR provides a legally binding agreement for the coordination of the management of public health emergencies of international concern. It is designed to improve the capacity of all countries to detect, assess, notify, and respond to public health threats. To date, 194 nations have become parties to the IHR, which requires a plan of action to be developed to ensure that these core capacities for biosurveillance and response were functioning by 2012. The result will be a significant contribution to global public health security.

Countries that are parties to the regulations have two years to assess their capacity and to develop national action plans, followed by three years to meet the requirements of the regulations regarding their national surveillance and response systems as well as the requirements at designated airports, seaports, and certain ground crossings (a two-year extension may be obtained, and, in exceptional circumstances, an additional extension could be granted, not exceeding two years).

So while the BWC is not the only international forum for a dialogue on biodefense issues, it clearly plays an important role and would become central in the event of a state or terrorist BW incident. If one considers a scenario whereby a significant outbreak of a contagious disease such as Rift Valley Fever (RFV) that is endemic in parts of Africa was to occur, we would expect to have reporting through multiple reporting chains. The WHO would receive reports on the human spread of disease. The OIE might likely receive reports based on the spread of the disease in the animal population. The FAO reporting would focus on providing information on mitigating the spread of the disease. The BWC channels would likely activate if the disease was seen to deviate from the normal pattern, triggering both a potential investigation and certainly an annual reporting requirement as part of the BWC CBMs.

4. ...has an important economic dimension.

Some 20 percent of the U.S. Gross National Product (GNP) comes from medicine, public health, biotechnology, and agriculture, which is directly related to the BWC. Arguably, this is a difficult figure to calculate, but approximately $3.5 trillion per year of the United States' $14.3 trillion economy is related to these fields. We can expect the percentage contribution from the life sciences and related fields to increase in this Age of Biotechnology.

Advances in existing fields of research and even new emerging areas are seeing a technological revolution that even exceeds what we have seen in the Information Age. In a comparison of the fields of information technology and biotechnology, consider that the rate of change as articulated in Moore's Law,[33] which states the number of transistors that can be placed on an integrated circuit doubles approximately every two years, is eclipsed by Carlson's Curve, which links the growth in biosequencers and biosynthesizers to advances in biotechnology. In his work, Carlson demonstrates that the costs of reading and writing new genes and genomes are falling by a factor of two every eighteen to twenty-four months, and productivity in reading and writing is independently doubling at a similar rate.[34]

One blog site provides "Biotech Trends in 2012," listing ten changes expected in the industry:[35]

- Social Media in Biotech and Health Care
- Synthetic Biology
- Direct-to-Consumer Genetic Tests
- Comparative Effectiveness and Personalized Medicine
- Patent Reform
- Commercialization by Nonprofit Foundations
- Electronic Medical Records
- Approvals for Genetically Engineered (GE) Animals
- Follow-Up Biologics
- Shifting IP Constituencies: Innovator Pharma Buys Generics and Asia Turns to Innovation

Each of these changes has significant economic implications in the United States and throughout the world. Some have the potential to change existing fields dramatically, such as the use of social media in health care and the use of electronic medical records. These have the potential for increased efficiencies and the reduction of costs in the delivery of health care. Others, such as synthetic biology, direct-to-consumer genetic tests, personalized medicine, and genetically engineered animals are emerging areas of research and opportunity that are growing exponentially. Each is so early in its life cycle that the full promise of these technologies has yet to be felt, but the potential is clearly there for groundbreaking and even life-altering outcomes.

Also noteworthy are the global shifts in biotechnical expertise and capacity that are resulting in a growth in capacity in Asia. This is not to say that the United States is no longer an innovation leader in biotechnology, but rather that others such as China, India, and Singapore are emerging as coleaders on biotech innovation and in some cases eclipsing us. The trend is important for several reasons. The economic incentive to lead biotech innovation is clear. There is also the concern that some of the development might be occurring without the benefit of proper containment for conducting experiments and that laws, regulations, and policies may not be complementary to those in the United States. Both of these shortfalls should be of concern as the economic consequences of a bioevent, whether naturally occurring or man-made, could be devastating.

Agriculture and the food industry contribute more than $1 trillion to the economy per year and one-sixth of the gross national product (GNP). One of every eight Americans works in the agricultural sector, including farmlands, feedlots, processing plants, warehouses, research facilities, factories for food preparation and packaging, and the national and global distribution network. The economic impact of an agricultural biological threat would include direct loss of crops, livestock, and assets; and secondary losses in upstream and downstream markets, lost export markets, significant price effects, and a reduction in economic growth. A single Foot and Mouth Disease (FMD) virus-infected cow in the United States would have a several billion dollar effect on the agricultural sector in the first days alone, shutting off exports and calling into question the safety of the U.S. livestock industry. Recovery would take years as the recent outbreaks in the United Kingdom and Japan have demonstrated. The 2001 U.K. outbreak caused 6.6 million animals to be destroyed and sixty suicides by U.K. farm owners; the economic impact was estimated at $10.7 to $11.7 billion USD. The spring 2010 FMD outbreak in Japan and South Korea was estimated to cost $10 billion to $50 billion by the end of the outbreak. By one estimate, a limited FMD outbreak on just ten farms in the United States could have a $2 billion financial impact.

FMD is not the only agricultural animal pathogen of concern. Others include rinderpest, which after a global eradication campaign saw the last confirmed case in 2001, classic swine fever, African swine fever, and Newcastle disease, to name a few. For plant pathogens, wheat rust, rice blast, and potato beetle have been and remain pathogens of concern.

We have also seen the impact with naturally occurring diseases such as influenza. Annually, we experience an influenza season that has both human and economic consequences. Some years the virus is more virulent; others it is more transmissible, and still others it is both more lethal and transmissible, the worst possible combination. Thankfully, recent seasonal influenzas have never again approached the 1918 influenza pandemic that represented the direst outcome. One study suggested that

> seasonal influenza epidemics from 1976 through 2000 were associated with substantial morbidity, including 120,000 annual hospitalizations and an annual average of 130,000 influenza-associated all-cause US deaths. These estimates also highlight the increased morbidity and mortality in older age groups and the pronounced variability in disease burden between seasons. For example, the 1984–1985 influenza season was severe, with an estimated 50,789 influenza-associated deaths, compared with a season of moderate severity (1978–1979 season), when only 7,608 influenza-associated deaths were estimated to have occurred.[36]

Aside from the direct loss of life, other losses include the costs of preparing for and treating seasonal influenza, the loss of productivity for those who have been infected and caregivers, and the losses caused by mitigation procedures such as closing schools. In this regard, the 2009 H1N1 virus had a $500 million economic impact.[37] The loss from the culling of animals based on the mislabeling of influenza as the swine flu resulted in the killing of over three hundred thousand pigs in Egypt alone.

While the Severe Acute Respiratory Syndrome (SARS) in 2003 infected only eight thousand people globally, the disease spread to thirty countries, and its effect on the global economy totaled $8 billion.[38] The SARS pandemic has also had an effect on the global health biosurveillance and response systems, as more states have come to understand the need to have situational awareness before, during, and after an outbreak. They have also developed greater reliance on international cooperation and response systems that support their publics and have quickly demonstrated competence in dealing with biological crises.

For both the FMD and SARS viruses, we have been addressing naturally occurring disease outbreaks. However, they also demonstrate the economic consequences that are possible from a deliberate attack. Previously, the White House estimates from a postulated medium release using Washington, D.C., as the target were that 1.9 to 3.4 million people would require antibiotic treatment, approximately 450,000 would be made ill, and 380,000 of those would die, and the city would need to be decontaminated at a cost of $1.8 trillion.[39]

While populations are likely to be understanding, at least in the early stages of an outbreak, a deliberate attack is more likely to raise the ire of the people, with questions such as: Why were we not protected? What is wrong with the government that failed to plan for this type of attack? And perhaps more tangibly, who will make the reparations for the economic losses? The other side of the argument—that made in particular by the Non-Aligned Movement (NAM)[40] and developing nations—highlights their assertion that the benefits of science and technology should be shared among BWC member nations. It argues for an economic distribution of biodefense capabilities.

The intellectual property issue also has implications for biodefense work that is being conducted with international partners. Prior to entering into agreements under the Nunn-Lugar Cooperative Threat Reduction's biological defense program, nations are demanding clauses be inserted into the agreements that protect their ability to capitalize on any benefits that accrue. For example, the pathogens that have been collected and catalogued are being considered as state property, with nations requiring right-of-return agreements for pathogens that leave their territory. This is largely due to a perceived economic benefit should the pathogen eventually be used as a new-age drug, included in a vaccine, or otherwise prove to be profitable.

Some developing nations see the perceived lack of sharing with respect to science and technology as a hindrance to their economic development and a direct violation of other states parties' obligations under the BWC. In contrast, developed nations with strong biotechnological capabilities are not currently nor are they likely in the future going to be willing to have an unfettered "sharing" of biotechnology. Clearly, the Convention does not require a complete sharing, as the language is high-level and aspirational rather than stated hard-and-fast requirements. Still, the NAM and some developing nations have interpreted the Convention to have far greater requirements in this area than many of the potential technology providers would be willing to give.

When one thinks about the BWC, it is highly unlikely that the first consideration would be the positive effect that the Convention could have with respect to management of disease and longevity. Yet this is precisely one of the outcomes that could result from the successful implementation of the BWC. The Convention has a role in the entire spectrum of biological threats from naturally occurring endemic disease to the deliberate use of BW. The fundamental changes to disease management and ultimately to longevity will have significant economic implications. More resources will be necessary should the full weight of the Convention be implemented and the benefits to combating diseases such as tuberculosis, malaria, and HIV/AIDS that kill six million people globally each year are realized. The societal and economic consequences of improvements in life expectancy will also be stunning. Aging populations will place greater stress on nations with declining working populations, higher demands for resources (including food and agriculture), and encroachment into naïve areas that will expose larger percentages of the population to new emerging infectious disease (EID).

5. ...is gaining more importance as the spectrum of biological threats grows.

History has demonstrated that research in biology, even when conducted without any military application in mind, may still contribute to the production of biological weapons. Indeed, people figured out how to intentionally spread illness long before naturalists discovered that germs caused disease.

At the time of entry into force of the BWC in 1975, the primary biological threats were seen to be from the deliberate use of BW from state actors coupled with concerns about naturally occurring disease. Today, the list of threats has increased threefold. We continue to have concerns about several states that are either thought to have a full BW program, those states whose intentions are not fully understood, and those states that may be conducting research and development on BW. While the list of states appears to be down from the twenty or so nations that were thought to have BW programs in the assessments in the 1980s and 1990s, there are still several states, including North Korea, Syria, and Iran, that are thought to have BW programs.

It is worth highlighting that there is considerable difficulty in determining which states have BW programs and their technical sophistication or their capacity in terms of types of pathogens, quantities, and levels of containment. The relatively small scale of a BW facility is problematic in this regard. For example, the cost to develop even a "large-scale" facility for producing a BW is quite modest. The *Militarily Critical Technologies List Part II: Weapons of Mass Destruction Technologies* "Biological Weapons Technology," published by the U.S. Department of Defense, says that a vaccine plant would cost approximately $50 million, but that a less elaborate fermentation plant could be built for about $10 million, clearly well within the funding of most countries. Given the advances in biotechnology today, a foregone conclusion is that virtually any state desiring to have a BW program could develop such a capability. While it may take a more advanced technological state to develop a highly efficient, pathogenic BW weapon and delivery system, even less technologically advanced nations should be able to develop a more rudimentary capability.

Concerning the potential for a deliberate use of BW, the *National Strategy for Countering Biological Threats* released by the White House in 2009 provides an important overview of the threat:

> The effective dissemination of a lethal biological agent within an unprotected population could place at risk the lives of hundreds of thousands of people.

> The unmitigated consequences of such an event could overwhelm our public health capabilities, potentially causing an untold number of deaths. The economic cost could exceed one trillion dollars for each such incident.[41]

In considering the state BW threat, estimates of scenarios and causalities vary widely, with many calling for ranges of a million people exposed and infected and several hundred thousand killed. Many of these assessments represent extrapolations from the former U.S. and Soviet programs.

We also continue to remain concerned about naturally occurring disease and even see this threat is increasing both in terms of the potential speed of an outbreak and the numbers of potential pathogens. Globalization and the accompanying movement of populations for travel and immigration are providing the perfect highway for the transmission of disease. The speed with which SARS spread across the globe provides ample evidence of this increase. As populations begin to infringe upon areas that heretofore have been uninhabited, people are interacting with wildlife—both plants and animals—with devastating consequences. The plethora of EIDs that have been discovered from 1940 to 2004 provides one data point concerning these interactions. In all likelihood, these diseases are not new; they are just new to humans. We can expect these trends to continue as humans continue to interact in new places such as the Amazon rain forests and African jungles. For naturally occurring outbreaks of exotic and/or emerging diseases, the exact threat remains unpredictable, but the impact to our security, economy, society, and health could be significant.

Adding to the threats of state BW and naturally occurring disease are threats caused by unintended consequences of experimentation, accidents, negligence, vandalism, and sabotage, and state or terrorist intentional use of BW. It is interesting that each of these reflect human behavior resulting from misuse of technology, either accidentally or deliberately. Stated another way, the Age of Biotechnology is placing more and greater capabilities into the hands of a wider array of actors.

The 2001 anthrax attacks in the United States—known as the Amerithrax attacks—provide an example of this misuse. While the likely perpetrator, Dr. Bruce Ivins, an employee of the United States Army Medical Research Institute for Infectious Diseases (USAMRIID), has been identified as a terrorist, one could also make the case that his behavior could be characterized as vandalism or sabotage as well. In the Ivins case, whether one labels the acts as terrorism, sabotage, or vandalism, it is clear that the threat we faced was the insider.

The misuse of science—either intentionally or inadvertently—is also a very real threat. Consider the case of a National Institutes of Health (NIH) study in which an ultravirulent strain of mousepox virus, a relation to the smallpox virus in humans, was developed. Dr. Mark Butler, a professor of molecular microbiology and immunology at St. Louis University, used a 2001 published Australian study in which a lethal variant of the mousepox virus was created by manipulating the wrong gene in the attempted development of a sterilization treatment for mice. Butler replicated the experiment hoping to identify an effective sterilization treatment, and he inadvertently developed a variant of the mousepox that was twice as lethal as the Australian variant. Of particular concern is that both of these viruses were created mistakenly by activating the wrong gene in the experiment. This anecdote also points to another concern—the ready availability of information that can be used inappropriately.

More recently, concerns have arisen about the two H5N1 studies conducted in 2011 that were attempting to better understand the mechanisms of transmissibility that would make the H5N1 virus both virulent and highly transmissible. Historically, the highly virulent H5N1 influenza virus has a mortality rate of 60 percent yet is not able to sustain human-to-human transmission. When news of the experiments were made public, significant concerns arose about the types of experiments as well as the conditions under which they were conducted. In both cases, the experiments were conducted appropriately given the guidance available today. However, some have continued to question whether an experiment that would be dealing with such a highly virulent and transmissible pathogen (in its genetically altered state) should not have been conducted in Biosafety Level 4 containment, the highest containment level. In this case, the accidental misuse of these capabilities and the potential for genetically engineered pathogens to "escape" has potentially devastating consequences.

Turning to the terrorist BW threat, the thresholds associated with terrorists successfully acquiring, processing, weaponizing, and conducting an attack have been significantly lowered. In the 1999 Project Bacchus, the Defense Threat Reduction Agency gave a six-person team one year, "seed stock," and $1.5 million to determine whether a small terrorist group could develop a BW weapon. The team was successful, although the pathogen they used was a harmless simulant. More recently Dr. Robert Carlson addressed the topic in discussions of DIY or Do-It-Yourself biology. He has provided a schematic of a biolab that he has constructed in his garage with instructions for the types and ranges of costs of the necessary equipment. His lab can be acquired for far less than the $1.5 million spent with the DTRA program.

Project Bacchus and the DIY bio examples both highlight the difficulties associated with attempting to detect small BW weapons programs. While a state program would have a modest signature assuming that capabilities for storage and loading of the munitions were part of the program, for a small program as one might expect for a terrorist actor, the footprint could be as small as a ten-foot-by-ten-foot room, with a modest amount of readily available equipment.

Of course, a note of caution is in order concerning a terrorist developing BW capability. Even if a terrorist is only able to develop and deploy a rudimentary weapon, he might still be able to achieve an important psychological effect. Therefore, if the weapon is not highly virulent so as to cause high mortality and morbidity but requires decontamination or frightens the population, then perhaps the terrorist will have achieved a degree of success.

The expansion of the potential range of threats has been recognized by several member state parties to the Convention as well as by the BWC Implementation Support Unit (ISU). Still, debate occurs among several of the member states as to the utility and application of the Convention. Some nations believe that the original purpose of the BWC as an agreement between states that limits state BW is the primary purpose. Others have taken a more progressive approach, believing that the full range of threats must be considered within the context of the BWC. This is more than an idle debate, as where a state stands on this issue has significant implications on the BWC verification debate.

6. ...provides a forum for coordinating preparedness and response capabilities against the spectrum of biological threats.

Biological weapons are inherently difficult from a preparedness and response standpoint. Difficult to detect. Difficult to treat. Difficult to analyze. Difficult to attribute.

Previously, the relation (or synergy) between the WHO's International Health Regulations (IHR) and the BWC articles were discussed. This is a positive outcome as we certainly do not want to design preparedness and response capabilities for deliberate use of BW and a separate set of capabilities for the remainder of the threat spectrum; this would be wasteful in terms of resources, but more than that, it would miss the more strategic point. In the initial stages of a bioevent there will likely be great uncertainty concerning whether the event is natural, accidental, or deliberate.

As the 2011 German *E. coli* (*Escherichia coli* O104:H4) outbreak demonstrated, the early epidemiological stages were focused on the question of the source. Was it a food-borne source? If so, was it naturally occurring or a deliberate infection? The strain of the bacteria was particularly virulent, with the illness characterized by bloody diarrhea and other serious complications, including hemolytic-uremic syndrome (HUS), a condition that requires urgent treatment. The toll from the outbreak in human terms in Germany was 3,785 cases and forty-five deaths as of July 27, only three months after the outbreak was discovered. Additionally, a handful of cases were reported in several countries, including Switzerland, Poland, the Netherlands, Sweden, Denmark, the United Kingdom, Canada, and the United States.

Another recent outbreak that dealt with a dangerous pathogen involved anthrax-contaminated heroin, which has resulted in a number of deaths and illnesses since 2009. The source of the contaminated heroin was Afghanistan, perhaps from contaminated soils or contact with infected animal skins.

One can imagine the differences in the investigation if in the early stages, the outbreak was assessed to have been intentional. While this remained in the realm of the international (through the WHO), national, provincial, and local public health authorities, a deliberate attack could have resulted in the triggering of the BWC articles that call for providing assistance, conducting assistance, and the sharing of science and technology. The public health authorities would have continued to be in charge of the medical outcomes, but the BWC could have played a role as well.

At this point, we must acknowledge that while the BWC calls for international cooperation, it lacks the current mechanisms to manage, coordinate, or even support such a response. The three-person Implementation Support Unit (ISU) certainly lacks the staff to do little more than collecting and posting the annual national confidence-building measure submissions, some facilitating of activities, and the coordinating for upcoming BWC meetings. It would be left to the individual member states to interpret and determine their individual support for affected nations.

Using the example above, if the outbreak was assessed to have been intentional, the resulting investigation would have fallen to existing organizations such as the United Nations Security Council, INTERPOL, regional governmental organizations, partner nations, and nongovernmental organizations, for example. Each of the standing organizations, whether governmental or nongovernmental, has specific capabilities that could be contributed to support the outbreak response. INTERPOL would have an important role in the investigation, forensics, and attribution. Individual nations with specialized capabilities such as the United States with its National Biodefense Analysis and Countermeasures Center (NBACC) have specific forensics capabilities developed in the aftermath of the 2001 Amerithrax attacks. None of the support could be "directed" under the BWC or by any other international mechanisms, but coalitions of the willing could organize to support affected nations just as with other large-scale disasters, such as the 2011 Japan earthquake, tsunami, and nuclear accident.

There are other impediments to a rapid response. First, no international biosurveillance system exists that considers the bioenvironment holistically. The One Health concept provides the strategic underpinnings; however, it is not supported by linked systems that collect data and analyze and disseminate warnings and guidance in real time. This criticism is shared by most nations, as few have real-time reporting systems for human health data and even fewer have combined systems that collate information on human, plant, animal, and climate data.

Second, even treating those affected by a significant bioevent could be difficult. Do nations share their medical countermeasures globally, or do they hold them in reserve for their own populations that might become infected? This scenario played out in Exercise Atlantic Storm, which postulated a deliberate release of *Variola major*, the virus that causes smallpox.[42] In one telling moment, Madeleine Albright, former Secretary of State and playing the role of the president of the United States in the exercise, was faced with the question of whether to release stocks of vaccine from the U.S. stockpile and, if so, how much? What about investigational new drugs (INDs) that are issued under Emergency Use Authorization (EUA) that might be given in a crisis but have not been fully licensed; should there be adverse side effects, what would be the liability?

Analyzing and attributing is also complex, as the Soviet examples of the 1979 Sverdlovsk anthrax accidental release and the yellow rain incidents in the 1970s in Laos and Cambodia demonstrated. Both investigations required lengthy periods of data collection, analysis, and discussion to come to a conclusion. In these two cases, we found that allegations in the Sverdlovsk case were proven, while the yellow rain incidents were judged to not be deliberate biological incidents.[43] While the state of bioforensics has greatly improved since the 1970s, even with these improvements, determination of the source of a bioevent is likely to be a challenge.

While just having a forum for coordinating preparedness and response capabilities against the spectrum of biological threats is important, we must acknowledge that the BWC provides little more than moral support on this issue. In this regard, much is left to the member nations to determine the level of support to be provided.

7. ...provides direct linkages to international security mechanisms.

One might ask why it is necessary to have special international security mechanisms for BW, as other mechanisms exist through the United Nations or other regional security forums to deal with such an unlikely event. Certainly this is true, but it is precisely the low-probability, high-consequence nature of BW that makes having a specific forum so vital.

Biological warfare is potentially the greatest danger governments and societies are facing, yet it is also generally regarded as not particularly likely. The 2001 Amerithrax attacks resulted in a fifteenfold increase in funding levels in the United States as a result and the increased visibility concerning weapons of mass destruction, in general, and BW, in particular, and in such reports as the *Commission on the Prevention of Weapons of Mass Destruction Proliferation and Terrorism* (commonly known as the Graham/Talent WMD Commission) biological issues have seen a dramatic increase in visibility. Indeed, on the issue of BW, our collective consciousness has been raised, presenting us with both opportunities and risks.

In contemplating BW, some basic "truths" govern our thinking. First, BW can be unpredictable. They are derived from biological material and therefore are highly sensitive to the environment. Treated properly with the right nutrients, pathogens can thrive. In the wrong environment, they are likely to rapidly decay and have far less than the intended effect. Second, they can have disease-causing effects over long distances. The tests from the former offensive programs—in particular those of the Soviet Union and United States—have demonstrated this potential. Much of what we know comes from testing and studies rather than an operational use of BW. Third, history provides little direct evidence of the potential of BW. Here, we must acknowledge that the history of BW use has been sparse, with few anecdotes on which to base the discussions. Inevitably, the discussion always turns to the very few events of biological warfare that have occurred in the "modern" era, that period of time from World War II to the present.

The UN provides a variety of international mechanisms for bringing forward security issues of concern. The UN Security Council (UNSC) is the central body for "managing" the international security system; it has five permanent members and a rotating membership for the ten other seats on the Council. The functions and powers of the Security Council include maintaining international peace and security in accordance with the principles and purposes of the United Nations; investigating any dispute or situation that might lead to international friction; recommending methods of adjusting such disputes or the terms of settlement; formulating plans for the establishment of a system to regulate armaments; determining the existence of a threat to the peace or act of aggression and to recommend what action should be taken; calling on members to apply economic sanctions and other measures not involving the use of force to prevent or stop aggression; and taking military action against an aggressor, as well as several other stated functions.[44]

Clearly, biological issues could be brought to the Security Council without the BWC, yet it is precisely through the BWC that the underpinnings of the biological issue are understood and managed. As with other arms control treaties and conventions that address specific behaviors or types of weapons, expanding on the issue both defines and provides an international legal and moral basis for the actions that can be taken by the UNSC and the full UN. Therefore, in the same way that other WMD are called out within international conventions and treaties, BW requires a special forum that provides direct linkages to international and national security and defense mechanisms.

The UN system of international laws allows for greater definition to be applied to specific areas such as nonproliferation, counterproliferation, and consequence management. This is precisely what the BWC provides in the area of biological issues. Of course, that is not to say that no other mechanisms will or should overlap. For example, UNSC Resolution 1540 under Chapter VII of the United Nations Charter "affirms that the proliferation of nuclear, chemical and biological weapons and their means of delivery constitutes a threat to international peace and security" and requires that nations refrain from "developing, acquiring, manufacturing, possessing, transporting, transferring or using nuclear, chemical or biological weapons and their delivery systems."[45] Clear overlap exists between the BWC and UNSCR 1540, but they are reinforcing and complementary, and not in any way contradictory. The same can be true concerning resolutions pertaining to specific national WMD programs, such as UNSCR 1441 relating to Iraq. This 2002 UNSCR served as the basis for the overthrow of Saddam Hussein in the second Gulf War in Operation Iraqi Freedom.[46] These "overlaps" serve to amplify or reinforce the broader international law in an important manner, on many occasions providing a legal basis for taking action.

Likewise, laws, policies, and norms that emanate from these UN frameworks can lead to other supporting programs. Examples include such programs as the Proliferation Security Initiative (PSI) that seeks to counter the proliferation of WMD material through training and education, and the Australia Group, a coalition of like-minded nations that seek to limit the proliferation of dual-use equipment, information, and material.

The importance of the BWC bringing together experts to deal with nonproliferation, counterproliferation, and consequence management for biological issues cannot be overstated. The time to consider these mechanisms is prior to the proliferation of dangerous biological capabilities. To do so after the proliferation has occurred or after an attack is a reactive strategy that places the international community at great risk. The time to prepare is now, and the BWC is the forum that can be used to drive this collaboration and coordination.

8. ...is an arms control agreement that relates directly to public health, the environment, food security, and biodiversity.

The BWC is unique in the degree to which it has applicability to the daily lives of citizens and the arcane world of arms control and national security. The "big daddys" of arms control have been the nuclear agreements signed between the United States and the Soviet Union. Legions of arms control experts cut their teeth talking about throw weights, multiple integrated reentry vehicles (MIRVs), and circular error probable (CEP). Few people, comparatively, worked on nuclear weapons development, although many policymakers fancied themselves as nuclear arms control experts. With the NPT, 194 nations signed the agreement, but far fewer had any nuclear capability. It was essentially a legally binding commitment to not proliferate nuclear technologies. These were important agreements. They reduced East-West tensions during the Cold War and limited the proliferation of dangerous nuclear weapons capabilities and the potential for conflicts to become nuclear. These were all very important outcomes. But on a daily basis, these agreements had little *direct* influence on our lives. Some, who follow the national security issues of the day, might have felt more direct influence from these agreements, but largely populations went about their daily business oblivious to the particulars of the negotiations or their outcomes. They depended on national security experts and negotiators to protect them from these threats.

Other arms control treaties such as the CWC or the United Nations Conventions on the Law of the Sea (UNCLOS) have direct influence on our daily lives, but they have considerably less visibility than nuclear weapons treaties. For example, the CWC pertains to chemical weapons and requires the destruction of all remaining stocks of these dangerous weapons. It also has requirements for the control of dangerous precursor chemicals and a verification regime. The CWC breaks down chemicals into Schedules 1 through 3 based upon scale (i.e., amount or quantity) of usage and potential to be used for chemical weapons. The Convention also has provisions that allow for plants that manufacture more than thirty tons per year to be declared and potentially inspected. Additionally, the CWC provides restrictions on export to countries that are not CWC signatories. The use of chemicals has become pervasive across the globe in everything from plastics to medicines. UNCLOS defines the rights and responsibilities of nations in their use of the world's oceans by establishing guidelines for businesses, the environment, and the management of marine natural resources. It has become more important in this age of globalization in managing activities at sea, including shipping and commerce as well as the exploration in places such as the Arctic, and discussing territorial claims.

The argument advanced here is not that these other treaties and conventions are not important, but rather that in terms of the relationship to the everyday health and well-being of societies and populations, the BWC, with its linkages to public health, biotechnology, agriculture, medicine, and health care is or should be several orders of magnitude more important than other arms control treaties. Decisions made in the BWC actually have the potential to both positively and negatively influence our lives in dramatic fashion, yet few people outside of the communities of interest directly related to the Convention have any realization of this importance. Even within security and defense communities, the Convention is treated as an afterthought, with review conferences held every five years, an intersessional process that is for information only, and delegations that largely perform double duties as representatives to the Conference on Disarmament (CD) and the BWC. Admittedly, the CD-BWC relationship is a natural fit in that the CD served as a basis for the development of the BWC, but in this Age of Biotechnology, when change is occurring exponentially, when the number of challenges and opportunities is growing accordingly, and when the economic dimensions of the BWC have come into such clear focus, we have failed to provide the necessary emphasis to the Convention.

To demonstrate the linkages between BW preparedness and public health, one analysis depicted BW preparedness as including biodefense, forensics, biosecurity, scenario training, and interinstitutional coordination; shared tasks including disease surveillance, detection, and disease outbreak response activities; and public health having a single task, noncommunicable disease activities.[47] A similar analysis has been depicted within the U.S. Department of Defense "Biosurveillance Roles, Missions and Functions" document dated November 2011. In short, the concept of the securitization of health or the realization that health and security are inextricability linked has gained wide acceptance.

Another growing realization of the important relationship between health and security concerns the notion that one of the best preventative measures in the event of a BW attack is the health of the population. A potential well-nourished individual with a robust immune system has a greater capacity to ward off infection, either naturally occurring or man-made. On the other hand, those with underlying health conditions will be more susceptible to the effects of a BW attack or a naturally occurring pandemic.

As further proof of the emerging linkages between health and security, at the highest levels of the biodefense issue, the linkages have been identified and greater collaboration is occurring between the WHO, FAO, and OIE and the BWC, as was discussed previously, but still these interactions remain at a very thin crust of the experts from these communities.

But why is it important that this understanding be brought to lower levels? Because when you go into an emergency room, a clinic, or to your health care provider with a fever and runny nose, you want to make sure that your health care provider understands the differences between the symptoms associated with seasonal influenza and a BW attack. Today, most clinicians are not trained in this manner, and diseases such as anthrax, plague, and dengue fever—that have a BW potential—are so uncommon that when victims arrive at clinics, emergency rooms, and health care providers, frequent initial misdiagnosis occurs. An analogy describing this issue is that when doctors hear hoof beats, they have been trained to think horses, not zebras; however, in this globalized world in which naturally occurring disease can travel around the world with great speed and terrorists have demonstrated the desire if not necessarily the expertise to acquire and use biological weapons, health care providers must broaden their repertoire to include both horses and zebras.

Biological weapons can have a deleterious effect on the environment in several ways. First, in a primary sense, biotechnology has advanced to the point where we are able to use microbes to clean up spills, wastes, and other contaminants. But this technology also provides the capacity to do damage to the environment by misapplying the benefits for nefarious purposes. The potential clearly exists for using these same technologies to degrade weapons systems, truck tires, and enemy fuel supplies. As a secondary impact, we can consider the unintended effects. Certain biological weapons including spore-forming organisms such as *Bacillus anthracis*, the causative agent for anthrax, and even other pathogens if properly prepared and deployed could have a significant impact on gaining access to areas following an attack. Such was the case of the British return to Gruinard Island off the coast of Northwest Scotland that had been used in the 1940s for open-air testing of anthrax. The decontamination effort took years and millions of gallons of seawater. Finally are the tertiary impacts associated with the environmental effects of BW. One of the questions the government would face following a biological attack would be what amount of decontamination is necessary prior to allowing people to reenter a space that had been exposed to a deadly pathogen. Of note, there is a technical answer to this as well as a psychological component; we do not know what it will take to make people feel comfortable returning to a location that had been attacked with a biological weapon.[48]

The use of BW can also have a severe, negative impact on food security through both direct and indirect attacks on the food supply. By direct attacks, one could postulate the use of antianimal and anticrop BW designed to cripple part of the economy. Previously, the impact on the $1 trillion per year agricultural sector was discussed in terms of the potential economic consequences of a BW attack. Attacks of this sort would have second-order social effects that could also be devastating to the workers in the food supply chain that goes from the farm to the table. The 2009 pandemic influenza that was originally identified as the swine flu pandemic demonstrated the potential for unintended consequences on the food supply chain. An example is in Egypt, where more than three hundred thousand pigs were culled in the weeks immediately following the announcement of the pandemic.[49]

In one sense, though, the same discussion regarding the importance of a healthy population can be made regarding the food supply chain. Great benefits accrue to nations that have secured and resilient food supply chains that have protection from either naturally occurring disease or deliberate attacks. Robust biosurveillance systems, new techniques for animal and plant farming that protect the food from the environment, and a rapid ability to get messages to populations have positive, dual-use benefits.

Finally, biodiversity is an important concept for protecting our environment. One account states, "Disease outbreaks among wildlife and human populations (especially indigenous populations in developing countries) 'could result in severe erosion of genetic diversity in local and regional populations of both wild and domestic animals, the extinction of endangered species and the extirpation of indigenous peoples and their cultures.'" An example of the potential impact on biodiversity is the introduction of small-pox into the Native American population in 1763. The effect on this naïve population resulted in an epidemic that devastated many of the Indian tribes.[50]

9. ...relates to dual-use capabilities in a way that no other arms control treaty does.

> The great achievements of molecular biology and genetics over the last 50 years have produced advances in agriculture and industrial processes and have revolutionized the practice of medicine. The very technologies that fueled these benefits to society, however, pose a potential risk as well—the possibility that these technologies could also be used to create the next generation of biological weapons. Biotechnology represents a "dual use" dilemma in which the same technologies can be used legitimately for human betterment and misused for bioterrorism.[51]
> —The Fink Report, "Biotechnology Research in an Age of Terrorism"

Biotechnology and biological warfare are two sides of the same coin. They use the same techniques, materials, facilities, and experts. The differences between biotechnology and biowarfare are measured not in capabilities, but rather through the intent of those actors conducting the research. On one side of the coin are activities in fields such as medicine, vaccine and drug production, food safety, and agriculture for the benefit of humankind; on the other side are those activities that would use these same components to develop deliberate disease-causing weapons. The obvious example is gene therapy, which can be used in the treatment of disease for the repair of defective genes. This same technology has great application in the manipulating of genes in a pathogen to change the pathogenicity or stabilization characteristics. Therefore, beginning from a common starting point, one can arrive at very different outcomes with very little to distinguish the process.

Another example comes at the intersection of molecular biology, immunology, and tumor genetics, which has allowed extraordinary advances in vital areas including vaccine therapy, cancer therapy, drug therapy, and immunotherapy. The science behind these advances is the use of virus to ferry foreign genes to the body, in the cases of legitimate use, to provide treatments, therapeutics, and vaccines. The same technology could be used to deliver deadly pathogens with even greater efficiency.[52]

Because the BWC relates to activities that fall within this definition of dual use, the potential exists to inadvertently affect legitimate research while attempting to close a biological proliferation window. Therefore, we must pay particular attention to ensure that limitations imposed through the BWC do not tip the tenuous balance between limiting dangerous activities and those that are to the great benefit of humankind. Unfortunately, the balance is becoming more difficult to discern in this Age of Biotechnology.

The changes in biotechnology are occurring at such a rapid rate that laws, regulations, and policies are being outpaced. Two potential outcomes result from such a situation. Either experimentation is being done or advances made with little or no structure with respect to legal, moral, and ethical requirements, or alternatively when limits are developed, they are out of date and therefore either irrelevant or harmful to the legitimate pursuit of science. Neither is desirable or acceptable. An example of the first case is the U.S. government-supported H5N1 influenza transmissibility studies conducted by Dr. Yoshihiro Kawaoka at the University of Wisconsin and Dr. Ron Fouchier at Erasmus Medical Center in The Netherlands. The questions came down to whether the articles should be published, was there any way to stop or influence what is published, and whether the studies were legitimate scientific pursuits. In the case of the last question, some argued that the experiments were not legitimate as they looked to increase the transmissibility of a pathogen, while others questioned whether proper containment techniques had been used. Ultimately, the National Science Advisory Board for Biosecurity (NSABB) recommended and the Department of Health and Human Services (HHS) agreed that the articles should be allowed to be published once the appropriate factual and editorial corrections were made.

Perhaps the more fundamental questions that should have been asked were whether the experiments should have been conducted at all. Was the anticipated gain such that the risks were outweighed? Were there other, more appropriate ways to conduct the experiments? Were the experiments conducted in the appropriate level of containment?

Still, the debate points to the very difficulty that we are facing. We must have some collective understanding of what experiments should be allowed and those that have the greatest potential for misuse or even accidental release with potentially devastating outcomes. In this decision, a risk-benefit analysis provides a basis for deciding on whether an experiment should be conducted. It is based on what has become known as the NSABB seven experiments of concern:[53]

1. Would it demonstrate how to render a vaccine ineffective?
2. Would it confer resistance to therapeutically useful antibiotics or antiviral agents?
3. Would it enhance the virulence of a pathogen or render a nonpathogen virulent?
4. Would it increase transmissibility of a pathogen?
5. Would it alter the host range of a pathogen?
6. Would it enable the evasion of diagnostic/detection tools?

7. Would it enable the weaponization of a biological agent or toxin?

The guidance states that experiments that would fall within the "experiments of concern" should be only conducted under approved conditions, with appropriate oversight and in the proper environments with trained personnel. All of this seems reasonable, but one must consider that no line exists that delimits legitimate research from nefarious or unsafe activities. Judgments concerning research must be made on a case-by-case basis in consideration of the totality of the experiment to be conducted.

The point of this discussion is to highlight this delicate balance and ensure that legitimate research can be conducted, while dangerous research will be appropriately controlled or regulated. So in the example of the H5N1, some might argue that the experiments should not have been allowed. This position would need to be balanced with the need for legitimate research that could assist in the understanding of transmissibility and the paucity of the mutations required for the current low-transmissibility virus to mutate and become highly transmissible while still maintaining its virulence. For influenza specialists and public health professionals, these are important questions. On the other hand, the techniques the researchers used provide insights into the engineering of a deadly pathogen.

To bring this argument to a close, it is worth reiterating that no other arms control treaties concerned with weapons of mass destruction capabilities have the degree of dual-use concern as the BWC; this is based on two important factors: availability and consequence. Nuclear capabilities are the purview of a select group of scientists and engineers. Further, nuclear material is not readily available and requires a high degree of processing to obtain. Chemical capabilities, while widely used and contributing to advances in other areas and industries, present far less of a dual-use concern. As with nuclear, they are more tightly controlled in the manufacturing processes and therefore are less readily available. Biological capabilities relate to some 20 percent of our economy and therefore are inherently dual use. Based on consequence, nuclear is of concern, but here the issue comes down to such limited availability due to the control regimes. For chemical, any use would be of tactical nature, with less chance of high mortality and morbidity unless massive quantities were used. The opposite is true of biological capabilities, which have been demonstrated to have the potential for strategic consequences even in small quantities and are readily available.

It is in the area of biotechnology that the dual-use implications have been the greatest. We should expect this trend to continue in this Age of Biotechnology. To provide a relevant BWC example of this dual-use issue, one of the concerns about establishing a verification protocol surrounded the potential for the regime to contribute to industrial espionage while providing no useful transparency or assurance of BWC compliance.

10. ...has responsibilities for implementation that run the gambit from international organizations to the individual.

Implementation of the BWC is unlike implementation of any other arms control agreement. Inherent in most arms control treaties is the exclusivity of the state in managing the elements of the negotiation. In nuclear arms control, the state or perhaps a limited number of commercial corporations under strict guidelines and subject to inspections through an international body, in this case the International Atomic Energy Agency (IAEA), control behavior, monitor safety, and assess compliance with existing laws, policies, regulations, and treaties. While individuals have a role in this specialized nuclear arms control arena, they are highly trained engineers and scientists and not a broad cross-section of the population. The same can be said for the CWC, which provides limitations on chemical weapons.

For the BWC, implementation responsibilities cannot be levied on a single international organization and a small number of specially trained people. Member states have responsibilities for national implementation, including establishing laws, policies, and regulations for all components of the BWC—development, production, stockpiling, or otherwise acquiring or retaining of biological material—that have "no justification for prophylactic, protective or other peaceful purposes." The London-based Verification, Research, Training and Information Centre (VERTIC) provides statistics on the dire state of BWC national implementation, indicating that by some measures, state compliance is less than a third of the member states in key areas such as having laws preventing storage and transfer of dangerous pathogens.[54]

Member states also have responsibilities in such areas covered by the Convention as biosurveillance, disease reporting and response, and cooperation in preparedness activities. More on the exact nature of the Convention's national responsibilities will be provided in the article-by-article discussion in the next chapter. Suffice it to say, though, that the state responsibilities fall at this intersection between medical, public health, scientific discovery, law enforcement, security, and defense.

Many of the issues that fall within the purview of the BWC are crosscutting, requiring implementation at a number of levels simultaneously. Consider biosurveillance, where the International Health Regulations (IHR) established by the World Health Organization (WHO) and agreed to by 194 nations requires states to have biosurveillance capabilities and to share in real-time information on emerging infectious disease. This necessitates having national biosurveillance systems that can collect, analyze, and feed information to the WHO. It further implies that "feeder" systems exist at state and local public health levels that can provide information to the national authorities. Of course, public health officials are not health care providers with access to the original information on patients; their function is to collate and examine data on public health issues. Therefore, they rely on doctors, clinicians, and other health care providers for providing the source information. First responders, such as police, fire fighters, and emergency management services also have an important role in the biosurveillance effort, especially in the event of an attack. By providing information on populations that are affected, we can gain a better understanding of where a BW attack might have occurred. Finally, individual citizens have an important role in biosurveillance, particularly in understanding the differences between normal disease such as the common cold or seasonal influenza and something more serious that could be the result of a BW attack or require hospitalization. This is not to imply that our citizens should attempt to become their own doctors, but rather that they must have the situational awareness to understand when anomalous behavior is occurring and react accordingly. In fact a general theme emerging in the aftermath of 9/11 is the individual's responsibility for assisting governments at all levels to participate to a greater degree in their own security through greater awareness.

While the example above concerns biosurveillance, the same philosophy applies equally to other areas affected by the Convention, such as bioforensics. The BWC requires notification of anomalous disease events to include provisions for lodging complaints against a nation suspected of having perpetrated an attack and provides a basis for conducting the forensics associated with an investigation. These requirements and authorities emanate from the United Nations, WHO, INTERPOL, and BWC at the international level to state requirements to have proper capabilities for investigating anomalies within their territories. Again, below the state or national level, local authorities have a role, if not in the analysis of biological events that can require expensive and highly specialized tools then at least in the reporting and proper securing of sites, biological materials, and other capabilities that may have been used to perpetrate an attack. In the same way, individuals have a role in understanding if an anomaly has occurred, such as a "white powder" event; in this case, the goal would be for the person to notify competent authorities and remain clear of the area.

BWC implementation also requires that the shared responsibilities extend to such mundane areas as the use of antibiotics and antivirals. If these treatments are used incorrectly or not disposed of correctly, this increases the chance of resistance, which has the potential for long-term effects relating to the efficacy of these drugs.

Even in areas such as response, a shared responsibility exists in areas such as strategic communications. Governments must issue appropriate, clear guidance on actions to be taken in response to incidents across the spectrum from naturally occurring disease to a BW attack. Communities and individuals alike must follow the guidance to protect themselves, to limit the chance of additional exposures, and to assist in activities that support community resilience.

The potential for a strategic effect of a biological event, either naturally occurring or man-made, increases the likely scope, duration, and response requirements associated with such an event. Therefore, implementation of the BWC must be looked upon as a shared responsibility. To do less would only increase the risks to populations and places greater burdens on all segments of society.

CONCLUSIONS

In this globalized world and having entered into the Age of Biotechnology, we find the BWC at a strategic crossroads. Earlier we asked the question of whether the Convention should continue to muddle through, serving as an observer in the BW debate. Or will it evolve and adapt to meet the challenges posed by the rapid advances in biotechnology?

What are the threats we face from biological pathogens: state, nonstate, accidental release, inadvertent outcome of an experiment, naturally occurring disease? Does the BWC apply to all of these?

Is a verification protocol or compliance mechanism necessary, as many nations believe, or is it an anachronism of the arms control establishment? What does it even mean to "verify" when change is occurring with such regularity?

Is the establishment of the norm against the use of biological weapons through the BWC adequate to ensure that these types of dangerous weapons are never used?

So we are left with one final question regarding the BWC. If it is truly the most important arms control agreement of the twenty-first century, what can and should be done about it?

Chapter Four

Into The Abyss: The Articles of the Convention

THE BWC

Prior to the negotiation of the BWC, all weapons of mass destruction (WMD) arms control negotiations—since the 1925 Protocol for the Prohibition of the Use in War of Asphyxiating, Poisonous or Other Gases, and of Bacteriological Methods of Warfare—considered nuclear weapons. This was both an indication of the concern with which nuclear weapons were viewed as well as the belief that the 1925 agreement was adequate for limiting chemical and biological weapons. With the U.S. use of riot control agents and Agent Orange defoliant in Vietnam, the international debate livened concerning the use of chemical and biological agents as a form of war. The culmination of this debate was the BWC and the early language that would eventually lead to the CWC as unequivocal norms against the development, production, stockpiling, or acquiring or retaining of biological and chemical weapons, respectively.

In many respects, the BWC is a twentieth-century agreement in a twenty-first-century world. The original fifteen articles agreed to in 1972 form the basis of the BWC and have gone unchanged, with only relatively minor clarifications and new provisions introduced at regularly scheduled review conferences, or RevCons. These RevCons are held every five years and largely serve to reaffirm the articles through a laborious article-by-article review. Few new ideas surface during the conferences, and change in the BWC is measured at a glacial pace. Those new ideas that do are highly scrutinized and turned into consensus documents requiring little more than voluntary participation. The example of this phenomenon is the confidence-building measures (CBMs) that provide a format for reporting aspects of a state's biodefense program. The "required" format is specified but national participation is abysmal, with normally only seventy or so of the 168 member states making the required annual submissions.

The main interlocutors during the periodic RevCons are diplomats associated with the Conference on Disarmament (CD) located in Geneva, Switzerland. CD representatives normally lead the national delegations during the RevCons with augmentation from capitals for scientific and technical expertise. The RevCons are held in a large, circular, and cavernous room where representatives sit alphabetically by delegation. Observers are permitted to participate, although this role, too, is informational rather than as part of any decision-making process. The three-week RevCons are approached in a Scythian manner. The first week includes national statements of resolve and hope for reinvigoration of the Convention. During the second week, positions begin to become staked out and hardened; alliances are formed. The third week is dedicated to arriving at a consensus document that advances the BWC very little, if any.

Recently, the periodic RevCons have been augmented with annual meetings of state parties. This has created a more frequent interaction between member nations; however, these sessions are designed to serve as information exchanges only and not for decision making. Despite the increased frequency of BWC meetings with the inclusion of the annual MSPs and an intersessional process, the gaps between national positions show little signs of malleability. Additionally, there are also annual Meeting of Experts with technically focused presentations on topics such as outbreak investigations, new technical platforms for diagnosis, and biosurveillance; these meetings are held several months before the meeting of state parties (MSPs).

While the convention is a legally binding treaty, in many regards the BWC can best be described as a strategic statement of principles rather than a well-defined arms control treaty. There are no counting rules, no definitions of prohibited behavior, and few tangible requirements for member state parties. It resembles the preamble of an arms control treaty more so than the full treaty. It is compact, including about 1,700 words and fitting on less than five full pages of text.

It is interesting that in the final document, no verification arrangements were agreed upon. The differences of opinion on this issue go all the way back to the late 1960s and early 1970s when the treaty was drafted. For example, the Soviets expressed the desire to rely on "self-policing rather than on any plan of international control by inspection."[1] Others, in particular the European nations, expressed a strong preference for a traditional verification protocol.

In an internal assessment of strengths and weaknesses of the Convention, Dr. Piers Millet, one of the three members of the Implementation Support Unit (ISU) responsible for the BWC, provided a succinct summary. The

BWC strengths were identified as (1) a clear, comprehensive ban: no exceptions, few loopholes; (2) strong international norm, never publically challenged; and (3) future proof (so far . . .). On the other side of the ledger, the BWC weaknesses were identified as (1) no organization or implementing body; (2) no systemic monitoring of implementation or compliance; (3) no systemic assessment of needs or provision of assistance; (4) uneven national implementation; (5) no mechanism for investigating alleged violations; and (6) no direct coverage of bioterrorism; it was conceived to deal with state-based BW programs.[2]

ARTICLES OF THE CONVENTION

Fifteen articles form the basis of the Convention. They generally fall into three categories: enduring articles for compliance, one-time requirements, and housekeeping or administrative. The substance of the agreement is in the enduring articles. While many of the articles are quite compact, covering little more than a few lines, they have second- and third-order requirements that have significant implications for member states.

Article I: The BWC Statement of Purpose

The centerpiece of the Convention is Article I. It provides an absolute requirement for states to "never in any circumstances to develop, produce, stockpile or otherwise acquire or retain" biological agents or toxins that have no justification for prophylactic, protective, or other peaceful purposes. These statements are definitive, but herein lies the rub. None of the terms are defined. What constitutes development? How about production? If small quantities of highly pathogenic biologicals are produced and even made into an aerosol, does that always constitute a violation of Article I? Likewise, there are no definitions for the three Ps, prophylactic, protective, or peaceful purposes. The interpretation is left to the individual member states to determine if an activity being undertaken is permitted by or constitutes a violation of the Convention.

Additionally, Article I does prohibit "use" of BW weapons. Perhaps this is an irrelevant complaint given that a precursor for use would be the development, production, acquisition, or retention. Still it is a curious omission.

Another qualifier that is undefined and left to the interpretation of the member nation is what constitutes a type and quantity of biological agent that could be justified for prophylactic, protective, or other peaceful use. So what amount of pathogenic material would be reasonable to be stored or used in experimentation? If a nation were to retain one milliliter of pathogen in a liquid storage suspension, that quantity seems as if it could be easily justified to be retained for use in analysis and challenge tests in vaccine development or drug trials. How about one hundred milliliters? This quantity might be acceptable if the one hundred milliliters was broken down into one-milliliter vials, each containing a different strain. This would allow for testing against multiple strains and thus provide increased confidence in the vaccine or drug that is being tested.

So we have looked at the smaller quantities and determined there are appropriate uses for acquiring and retaining small amounts. How about the production of five thousand liters of *B. anthracis*—the bacteria that causes anthrax—in a fermenter? Again, the answer is it depends. Large quantities of biological material are required for vaccine production. So perhaps the quantity is justified. The case is certainly much easier to make if the strain of anthrax is one associated with vaccine development, rather than a highly pathogenic strain like those developed in the U.S. and Soviet programs. Still, the examples point to the overarching difficulty associated with implementation and ultimately with the verification of the Convention. If a verification regime were to be developed, clarification of all terminology would be an essential first step. However, even with clarification, the verification of Article I of the treaty would be highly subjective.

The lack of definitions for the key terms also hinders assessing whether an activity constitutes a violation of the BWC. Consider the case of a biologically derived toxin such as saxitoxin. Saxitoxin is an extremely potent neurotoxin that is secreted by algae. It is a naturally occurring pathogen found throughout the world in marine and freshwater areas where dinoflagellates and other similar algae live. Saxitoxin has been determined to be the cause of large marine die-offs and the associated "red tides." As with all toxins, there is a chemical formula that describes it. What if that formula is changed to something that does not occur in nature but still produces the same neurotoxic effect? Does it still count as a biological weapon? The classification in this case is not completely clear. However, the good news, of course, is that it would be prohibited under the CWC (assuming that the nation was a party to this Convention). Still, the example begins to suggest other issues, such as the development of genetically modified biological material or perhaps even laboratory-developed organisms that do not exist in nature. Would they be covered under the BWC? Recent discussions as part of the BWC suggest that there is general agreement that these areas would indeed be covered. However, does the simplicity of the BWC in its current form provide the richness to deal with the explosion in biotechnology that we are witnessing today?

The original conference in considering this question believed in the unequivocal application of the Convention and therefore that "toxins (proteinaceous and nonproteinaceous) of a microbial, animal or vegetable nature and their synthetically produced analogues are covered."[3] However, given advances in biotechnology, including the synthesis of living organisms as reported in 2009 by Craig Venter, the BWC will likely be challenged to keep pace.

The second clause of Article I prohibits the development, production, stockpiling, acquisition, or retention of "weapons, equipment or means of delivery designed to use such agents or toxins for hostile purposes or in armed conflict." During the U.S. and Soviet offensive BW programs, biological weapons delivery systems were tested that when used would allow for the release of the pathogen while minimizing the loss of biological material in an explosion. To this end, special bombs were developed that could be air dropped or ground launched that would release or eject smaller submunitions. In this way, the biological material would not be subjected to the forces of the initial explosion and heat, and the secondary munitions would allow for the controlled release of the material in a more even distribution by forming a cloud of biological agent.

Making a case for retaining these sorts of capabilities would be more difficult, however, not completely out of bounds. For example, from 1997 to 2000, the Clinton Administration authorized a project called Project Clear Vision with the stated goal of examining the dissemination characteristics of several types of bomblets used by the former Soviet Union.[4] In this case, the United States was heavily criticized for undertaking suspicious behavior of an offensive nature in violation of the BWC. The issue was compounded by the fact that the administration did not report this activity as part of the annual confidence-building measures (CBMs), thus increasing the suspicion that the activity was being kept secret and therefore related to an offensive capability. The U.S. position was that the activity was fully in keeping with the BWC, as the results were to be used to develop a better understanding of BW dissemination characteristics.[5]

In the case of Project Clear Vision, actual bomblet munitions were tested to determine dissemination efficacy. However, what if a high-capacity, industrial nozzle was being tested? Would that have made any difference to the perception? In fact, our experimentation from the former U.S. offensive program provides insight that suggests that spray nozzles from tanks affixed to jet aircraft was a very efficient method of dissemination.

The previous discussions about type, quantity, and dissemination methods suggest another important aspect of the BWC, the inherently dual-use nature of the Convention. There is a fine line between activity that is prohibited by the BWC and that which is allowed. Assessing where this line is and, therefore, whether an activity is allowed under the BWC requires a high degree of judgment and has the potential to lead to finger-pointing and recriminations, even in cases in which a preponderance of the evidence suggests that the activity is for prophylactic, protective, or other peaceful purposes.

This dual-use issue suggests an important test that must be met in justifying whether a behavior is or should be considered appropriate under the provisions of the BWC, the court of public opinion. Sometimes called the "Washington Post" test, it implies that whether a behavior is acceptable will be less about the facts of the case such as the intended use of the work and more about how the activity is perceived if or when it becomes public. As in the case of the failure to notify Project Clear Vision as part of the U.S. CBM submission, the perception of work being done as part of a secret offensive program was heightened by this omission.

In the final 2011 BWC RevCon document, a short synopsis of the agreements reached pertaining to Article I reflected a reaffirmation of the application of the provisions to "naturally or artificially created or altered microbial and other biological agents and toxins, as well as their components, regardless of their origin and method of production and whether they affect humans, animals or plants." Further, it reaffirmed that the article "applies to all scientific and technological developments in the life sciences" and that "open air release of pathogens or toxins harmful to humans, animals and plants" are not justified.[6] The most important aspect of the final declaration is the recognition that despite extraordinary advances in biotechnology and the life sciences, the BWC remains a viable and even an invaluable forum for addressing these critical issues.

Article II: Elimination of Current BW Programs

This article contains a one-time requirement of the Convention. It requires that "each State Party to this Convention undertakes to destroy, or to divert to peaceful purposes, as soon as possible but not later than nine months after the entry into force of the Convention, all agents, toxins, weapons, equipment and means of delivery specified in Article I of the Convention."

While the language seems straightforward, the implementation is far from direct. At the time of entry into force, states agreed to eliminate those stocks of biological material not deemed to be required for peaceful purposes. Herein we have another judgment to be made by individual member states that is far from absolute. One state may determine that retaining several hundred vials of multiple strains of *B. anthracis* is reasonable and even necessary, while another may believe that complete destruction would be warranted. So for completing the requirement of this article, several decisions must really be made. What pathogens are to be retained, for what purposes, and in what quantities?

Of course, should a state find it had inadvertently retained some of its previous stocks, Article II would assume relevance again. Such was the case with U.S. saxitoxin, where we believed that all such holdings had been destroyed in 1971, but it was discovered in 1975 that there were still some holdings at the Central Intelligence Agency (CIA). These were subsequently destroyed.[7] As history has demonstrated, not all nations have been honest in their original declarations of destruction. The insights gained into the former Soviet Union's (FSU's) BW program, which we have seen as a result of the Nunn-Lugar Cooperative Threat Reduction (CTR) program where previous biological weapons sites have been remediated in Russia, Azerbaijan, Kazakhstan, Ukraine, and Uzbekistan, to name a few, demonstrate this shortfall. The same can be said for Iraq, which despite being a member of the convention and declaring it had no BW weapons was found to have a covert program; more on this will be discussed later.

During the most recent review conference, the final language reaffirms that any state "ratifying or acceding to the Convention, the destruction or diversion to peaceful purposes specified in Article II would be completed upon accession to, or upon ratification of, the Convention."[8]

Notice also, no reporting requirements were established, such that nations that had BW programs would need to specify that offensive BW capabilities had been destroyed. It was left to the individual states to be "on their honor," so to speak. Years later, with the adoption of confidence-building measures (CBMs), nations could provide within their voluntary submissions the "declaration of past activities in offensive and/or defensive biological research and development programs since 1 January 1946."[9]

Articles III and IV: National Implementation

Articles III and IV address national implementation issues and are at the heart of the debate about limiting the free flow of scientific information. The tension arises from the requirement under national implementation to control biological material and the knowledge or other capabilities that might assist others in developing BW capabilities with requirements in Article X to encourage the peaceful uses of biological science and technology. These articles are a priority for the United States as they reflect the strong desire for member state parties to close potential proliferation windows by developing the appropriate laws, regulations, and policies to limit access to potential dual-use biotechnology. On the other hand, a strong perception exists, particularly among the developing nations and as expressed by the Non-Aligned Movement (NAM) at the RevCon, that Articles III and IV have the potential to limit scientific collaboration. The final language in the article-by-article review of Article III states, "Conference reiterates that States Parties should not use the provisions of this Article to impose restrictions and/or limitations on transfers for purposes consistent with the objectives and provisions of the Convention of scientific knowledge, technology, equipment and materials under Article X."

Article III requires that "each State Party to this Convention undertakes not to transfer to any recipient whatsoever, directly or indirectly, and not in any way to assist, encourage, or induce any State, group of States or international organizations to manufacture or otherwise acquire any of the agents, toxins, weapons, equipment or means of delivery specified in Article I of the Convention." The intent of this article is to limit proliferation. No member state should be providing offensive BW weapons or knowledge products to another state that could be used for nefarious purposes.

On the other hand, in practice, this article could have a deleterious effect on legitimate and even highly desirable international cooperation in the life sciences. Imagine a situation where based on our export control requirements scientists from two nations are collaborating on a scientific issue and sharing information and biological material during experimentation, which could be made illegal or become a violation of the BWC when the material is "transferred" between scientists in the lab. Perhaps this is an extreme example; however, a similar situation has occurred based on interpretation of Department of Defense regulations that deal with these issues.

But this limitation has broader implications that must be considered within the dual-use context previously described and the meteoric changes occurring within the biotechnology arena. Great difficulty exists in trying to limit transfer of dual-use capabilities that have equally important uses in developing biotechnical solutions and offensive BW. Consider the example of aerosol-delivered insulin, which is now available in a small nasal-spray device. The same capabilities for stabilization, microencapsulation, and aerosol delivery of insulin have direct application to developing offensive BW weapons.

Additionally, in an ironic manner, this article also provides, or at least can be seen to provide, conflicting guidance with the provisions of Article X, which calls for avoiding "hampering the economic or technological development" and the "fullest possible exchange of equipment, materials and scientific and technological information for the use of bacteriological (biological) agents and toxins for peaceful purposes."

Of course, the reference to Article I should allay the concerns of nations desiring to share dual-use capabilities with other nations, but still the language of the Convention is not clear on this point and requires a degree of interpretation. Does a nation, or perhaps even an international corporation, risk being tagged with violating the BWC because there is a sharing of knowledge products, microbial material, and specialized equipment that has equal application for use in medicine, public health, biotechnology, and offensive BW?

Article IV is most appropriately referred to as national implementation and states, "Each State Party to this Convention shall, in accordance with its constitutional processes, take any necessary measures to prohibit and prevent the development, production, stockpiling, acquisition, or retention of the agents, toxins, weapons, equipment and means of delivery specified in Article I of the Convention, within the territory of such State, under its jurisdiction or under its control anywhere."

For an international treaty, this article is unique and forms the basis for national implementation measures. Its uniqueness stems from the understanding that the BWC is not like other arms control treaties. The scope and reach, the dual-use aspects of the Convention, and the economic implications of the treaty combine to make the establishment of national implementation laws, regulations, and policies essential to successful national and international implementation.

In addition to laws, regulations, and policies, state parties should also have complementary codes of ethics and training programs to ensure that scientists are adhering to established standards and norms. The point would be to develop a culture among life scientists that reinforces the principles of biosafety and biosecurity. National compliance systems must also include periodic inspections of facilities to ensure proper standards of conduct are being followed.

For all of its importance to BWC implementation, it is one of the least well-implemented articles. Given the uniqueness of the BWC and its pervasiveness within society, one would expect a unique system of monitoring and enforcing compliance that begins with national capacities to ensure the proper use of the life sciences. An important foundation would consist of specific laws, regulations, and policies to monitor research with dangerous pathogens; import, export, and transit controls for biological materials; and protection for facilities where these dangerous materials are stored. Today, this largely does not exist in most countries. In fact, one assessment found that less than 25 percent of eighty-seven nations surveyed considered developing, stockpiling, and storing biological weapons a crime. The surveyed group fared slightly better in export control, with just over one-third of the nations having some measures; however, we should not be encouraged by this laissez-faire approach to national legislation for securing dangerous biological material. Facilities where work with these especially dangerous pathogens is conducted should have the appropriate security systems to prevent theft or loss of dual-use material. In the absence of these control measures, security of these facilities and the pathogens stored inside will remain questionable.[10]

National-level implementation also includes the development of public health systems such as biosurveillance, diagnostic capabilities, and treatment protocols that have an important role in early identification, the proper care of affected people, and the mitigation of threats. In this regard, great synergy exists between the provisions of the BWC and the World Health Organization's (WHO) 2005 International Health Regulation (IHR) agreement to develop national biosurveillance and response systems by 2012 and 2014, respectively.

Given the extensive overlap between medical, public health, biotechnology, and the BWC, national implementation of the Convention is more than just an important link; it is absolutely imperative to the BWC.

Articles V and VI: Mechanisms for Assuring International Compliance

Articles V and VI are related in that both provide mechanisms for international compliance issues and are enduring BWC articles, with Article V concerning bilateral and multilateral approaches to compliance and Article VI providing the authority to take alleged violations of the BWC to the UN Security Council.

Article V states, "The States Parties to this Convention undertake to consult one another and to cooperate in solving any problems which may arise in relation to the objective of, or in the application of the provisions of, the Convention. Consultation and cooperation pursuant to this article may also be undertaken through appropriate international procedures within the framework of the United Nations and in accordance with its Charter."

Article VI contains two provisions. The first allows for lodging a complaint against any state believed to be in violation of the BWC. It states, "Any State Party to this Convention which finds that any other State Party is acting in breach of obligations deriving from the provisions of the Convention may lodge a complaint with the Security Council of the United Nations. Such a complaint should include all possible evidence confirming its validity, as well as a request for its consideration by the Security Council."

The second component of Article VI provides direction for taking a complaint of violation of the BWC to the United Nations Security Council (UNSC). It states, "Each State Party to this Convention undertakes to cooperate in carrying out any investigation which the Security Council may initiate, in accordance with the provisions of the Charter of the United Nations, on the basis of the complaint received by the Council. The Security Council shall inform the States Parties to the Convention of the results of the investigation."

Taken together these two parts of this article, commonly referred to as the Secretary General's mechanism, essentially codifies a state's right to consult using already existing international provisions. While this might seem as somewhat feckless, the article can also be looked at more contextually as a very simplified verification mechanism or as a minimum, the overarching language for the development of a complete verification protocol. Still, requests for inspection must go through the Security Council, where they are subject to veto.

Throughout the history of the BWC, these two provisions have been used infrequently. One might think this odd given the history of state BW development, since the BWC entry into force in 1975 has seen some twenty nations develop various components of a state BW program. At the time of the BWC entering into force in 1975, four nations were believed to have BW, including the United States, the Soviet Union, China, and South Africa.[11] In 1989, Central Intelligence Agency Director William Webster announced that "at least 10 countries" were developing BW weapons.[12] In 1995, the Office of Technology Assessment at Senate hearings listed seventeen countries that were "suspected of manufacturing biological weapons."[13] In 2005, one source suggested that seven nations currently had programs, including China, Egypt, Iran, Israel, North Korea, Russia, and Syria.[14] Another indicated concern about "compliance for ten or so countries with the treaty and concerns about the biological weapons programs in a few countries that are not party to the treaty."[15] Obviously, the question of the actual number of BW-capable states was then and remains now a difficult one, as the data indicates. The lack of a viable BWC verification regime, coupled with the ease of masking a BW program, has contributed to this shortfall.

In considering the efficacy of the BWC provisions for gaining authorizations for conducting inspections, it is interesting to note that the two inspections conducted in Iraq (UNSCOM 1991–1996, UNMOVIC 2002–2003) were done under different UN Security Council resolutions—not under the BWC.

Also particularly noteworthy in the final 2011 RevCon language is the strong reference to the applicability of United Nations Security Council Resolution (UNSCR) 1540 that affirms support for arms control treaties that eliminate or prevent the proliferation of nuclear, chemical, or biological weapons.

So while Articles V and VI provide a mechanism for bringing forward complaints about other member state parties, in practice this has not occurred, even when there was a belief that a country maintained illicit BW programs. In the final document from the RevCon, the final declaration notes that "provisions of this Article have not been invoked."[16]

Articles VII and X: Support to Other Member State Parties

These articles have been the subject of intense debate before and during the 2011 RevCon. Prior to the RevCon during the intersessional meetings and at the annual meeting of states parties, states, especially members of the NAM, have strongly reiterated their desire to have greater access to biotechnology capabilities as called for in these articles. Developed nations have been more cautious with respect to these articles, not wanting to be coerced into a requirement to have to support the public health systems of other member state parties or have the requirement to share proprietary science and technology.

Article VII states, "Each State Party to this Convention undertakes to provide or support assistance, in accordance with the United Nations Charter, to any Party to the Convention which so requests, if the Security Council decides that such Party has been exposed to danger as a result of violation of the Convention."[17]

In the thirty-five-plus years since the BWC entered into force, member states parties and the BWC community in general have become more enlightened about the applicability of Article VII, realizing that the everyday preparedness and response capabilities that individual states undertake outside of the realm of the BWC in the fields of public health, animal and plant health, and emergency response planning do relate to the convention. Recognition now exists that satisfying the provisions of this article cannot just be considered at the time of a biological attack or incident, but rather must be part of all nations' national capacities. Therefore, it should come as no surprise that the final 2011 RevCon document should include reference to the World Health Organization (WHO), the World Organization for Animal Health (OIE), the Food and Agriculture Organization of the United Nations (FAO), and the International Plant Protection Convention (IPPC).[18]

Inherent in the modern interpretation of this article is the shared global responsibility to respond to the range of biological threats from naturally occurring endemic disease to a deliberate BW attack. Regardless of the source, the response must be a coordinated event with sharing across the global community. The Severe Acute Respiratory Syndrome (SARS) pandemic that begin in China in 2003 and eventually spread to thirty-two countries with an estimated eight thousand cases and 750 deaths demonstrates the global nature of disease.

So while the final document of the 2011 RevCon states that "the Conference notes with satisfaction that these provisions have not been invoked,"[19] this is really only part of the story. In fact, many nations have been working individually and as part of the larger global community to develop mechanisms for preparedness and response that would be critical in the event of a bioevent as well as having the benefit of improving the public health of humans, animals, and plants.

Of the two, Article VII is far less controversial and given the expansive, and growing, view expressed through the One Health concept, much of this effort is being incorporated into the daily global and national preparedness and response activities that are ongoing. That brings us to Article X.

Article X has two provisions. The first states that member nations "have the right to participate in, the fullest possible exchange of equipment, materials and scientific and technological information for the use of bacteriological (biological) agents and toxins for peaceful purposes" and that states should be able to "cooperate in contributing individually or together with other States or international organizations to the further development and application of scientific discoveries in the field of bacteriology (biology) for prevention of disease, or for other peaceful purposes."

This provision has been interpreted by the NAM to suggest that they should have unfettered access to science and technology related to the BWC. On the other side of the ledger, developed nations see this as infringing upon government and commercial research and development that has significant economic implications. The result has been a standoff, on which the sides have agreed to continue to have dialogue about the provisions for sharing science and technology.

Of course, the "complete" sharing of biotechnology is simply not feasible. Governments do not have a monopoly on biotechnology. In fact, most governments are not even the leaders within their sovereign territory with regard to these emerging capabilities. Take the United States as an example, where BWC-related scientific discovery is coming at the intersection between biotechnology, bioinformatics, and nanotechnology. In each of these three areas, industry and the scientific community are driving outcomes that are leading to exponential advances in key biotechnological capabilities. Additionally, concerns surrounding intellectual property rights are important to governments and private industry, and many nations do have export controls that both lead to limitations on the sharing of biotechnology.

A study done in 2000 by the Federation of American Scientists (FAS) examined twelve of these key capabilities, concluding that the rate of growth was exceeding 400 percent per annum in each. For example, in the field of vaccine development, the rate of development from 1940 to 1970 doubled every five years. However, from 1970 to 1980, the rate increased fivefold such that the time to double the capabilities in the field of vaccines was one year. Over the twenty-year period from 1980 to 2000, the time to double in capability decreased to six months. The technologies that the vaccine field contributes toward are vaccine development, sensors, personal protection, and pathogen masking. Another field, DNA engineering, not even in existence until 1982, has doubled in capacity every six months until 2000. This area is critically important to a wide variety of biotechnical advances, including gene therapy, vaccine development and sensors, as well as ominously increasing the virulence of a pathogen. The same is true for encapsulation and stabilization, which have potential for enhancing personal protection and therapeutics as well as making BW weapons more effective and stable in the environment.[20]

An article by Rob Carlson describes this growth and compares it to progress in information technology in what has become known as Carlson's Curve. In his formulation, he documents the growth in biosequencers and biosynthesizers as being exponential and even greater than that being experienced in information technology.[21] Carlson's work is also noteworthy for practices that have become known as DIY or Do-It-Yourself bio. He works out of his garage in a state-of-the-art lab that he has created from procuring secondhand biotechnical equipment.

So these advances are not due to a government-directed or sponsored effort to improve the life sciences. Rather, market forces, the aspiration for scientific discovery, and the desire to improve global public health drive them. Imagine the backlash that would ensue should the international community through an arms control agreement attempt to make the entire ongoing BWC-related discovery available to member state parties.

To drive home the point, consider that the medicine, public health, biotechnology, and related fields account in rough terms to almost 20 percent of the U.S. Gross National Product (GNP). This is both a consequential sum and not one that is reasonably captured in an international arms control treaty.

A second issue concerning the first provision of Article X concerns the potential for proliferation. In fact, Articles III and IV actually have provisions that would tend toward limiting exports of material, information, or equipment. Obviously, where the material is judged to be for "prophylactic, protective, or other peaceful purposes," export and scientific work would be allowed. Still, there is a definite concern about proliferation, especially given the dual-use nature of the biosciences.

To better understand this dual-use issue, consider the difficulties in discerning legitimate research from that done for malevolent purposes. In testimony to the U.S. Senate Committee on Governmental Affairs in May 1989, Colonel David Huxsoll, then director of U.S. Army Medical Research Institute of Infectious Diseases (USAMRIID) discussed the difficulties in early-stage work of differentiating between a vaccine development program and a BW weapons program. While there will be some telltale signs, it is really only in the later stages that intent becomes clear. Many of the steps are identical in the initial stages. Even in the later stages, differentiating between actions, equipment, and techniques for attenuating the pathogen to make it less virulent for a vaccine or more virulent for use as a weapon are difficult to discern. Similarities even exist in the testing of vaccines and weaponized material to determine effectiveness. One differentiation appears to be in the quantities of material produced; the vaccine development branch suggests that quantities would be small, while in the development of weapons, the quantities would need to be larger. However, even this is misleading, because in the manufacturing of vaccines, it is not uncommon to see large amounts of biological material that would be required for developing the capacity for mass vaccinations.[22]

Proliferation has been and will continue to be a significant concern in the application of Article X. One might ask, "Why should these limitations exist"? Consider the questions raised during the Dual Use Research of Concern (DURC) debate about the two H5N1 influenza manuscripts. The manuscripts from two different studies provided details concerning avian influenza transmissibility (spread from one animal or person to another) and pathogenesis (ability to cause disease). The manuscripts drew global attention about the need for, appropriateness of, and conditions under which "dual use research of concern" exist.[23]

The questions raised in this DURC discussion include: Were the experiments done in the proper containment? Were the manuscripts essentially recipes for developing a bioterror weapon? What was the risk-benefit of allowing the research to be conducted and the manuscripts to be published? While this debate was held outside of the BWC community, it demonstrates the degree to which these issues have relevance to the BWC as well as to the application of Article X. The DURC debate in which the BWC was silent also provides an example of where the convention could potentially have played an important role, but passed on doing so.

The second provision of Article X states that the "Convention shall be implemented in a manner designed to avoid hampering the economic or technological development of States Parties." Herein lies perhaps the aspect of Article X that creates the most controversy with the NAM.

The NAM continues to express the strong belief that failing to share the scientific and technology related to BWC fields has a deleterious effect on their economic well-being. The train of logic is that the failure of developing nations to not fully share the fruits of scientific discovery violates the spirit as well at the letter of the Convention. Of course, for the reasons discussed within the first part of the Article X provision, developed nations push back on providing this unfettered access.

During the 2006 RevCon, the NAM pushed for including science and technology in the intersessional process. The topic was also heavily discussed during the 2011 RevCon with the pledge to continue to have a dialogue on BWC-related technologies. Here the discussions become very nuanced. The NAM and several developing nations would like to gain access to biotechnology. The developed nations want to use the dialogue during the intersessional process to ensure that the BWC continues to remain relevant despite the significant advances occurring in these fields. In a sort of measurement process, the intent is to ensure that no changes in the life sciences would invalidate or obviate the BWC.

So the Article X debate continues.

Articles VIII, IX, XI, XII, XIII, XIV, and XV: "Housekeeping"

There is no intent to diminish the standing of these articles but rather to delineate that they serve a necessary, yet supporting, role within the Convention. Article VIII reinforces the importance of the Protocol for the Prohibition of the Use in War of Asphyxiating, Poisonous or Other Gases, and of Bacteriological Methods of Warfare, signed at Geneva on June 17, 1925. In fact, one can think of the 1925 protocol as a necessary and critical foundation of the BWC.

One might ask then, given the 1925 protocol, why to even have a separate agreement on biological weapons. This is an important question that begins with understanding the applicability or scope of the protocol. The 1925 agreement only applies to actions in war. Therefore, it does not have applicability with the type of behavior that was intended to be captured under the BWC. That is, the BWC was intended to prohibit the research, development, production, and stockpiling of these weapons. The potential for misuse or malicious use of biological agents was thought to be of such consequence that a separate international agreement would be necessary to "remove the threat of future use."[24]

Article IX was an important element of the negotiations of the BWC as it assisted in both delineating biological and chemical weapons and highlighting that negotiation on the elimination of chemical weapons would continue in a separate forum. The language of the article regarding potential chemical weapons provisions—to reach "early agreement on effective measures for the prohibition of their development, production and stockpiling and for their destruction, and on appropriate measures concerning equipment and means of delivery specifically designed for the production or use of chemical agents for weapons purposes"—is strikingly, yet not unexpectedly similar to the BWC language. Interestingly, the BWC entered into force in 1975 while the CWC did not enter into force until April 29, 1997; however, the CWC does have a robust verification annex. This "shortfall" in the BWC is seen as the single biggest issue by many of the member state parties. In another interesting contrast, the CWC lists 188 member nations, while the BWC only has 168 members.[25]

Article XI is one of the important articles that ensures the continued relevance of the Convention and provides a venue for states to make recommendations, allowing for proposing "amendments to this Convention."

Article XII provides for regular meetings of the BWC through a review conference held every five years for the purpose of reviewing "the operation of the Convention, with a view to assuring that the purposes of the preamble and the provisions of the Convention, including the provisions concerning negotiations on chemical weapons, are being realized. Such review shall take into account any new scientific and technological developments relevant to the Convention." While the necessity of having a regular forum for addressing BWC-related issues was understood, the framers of the Convention selected a time frame for review that was woefully inadequate for the task at hand. With changes in the life sciences that are occurring so rapidly, the member state parties came to realize that more frequent interaction was required, hence the development of the intersessional process that was initiated at the 2006 BWC RevCon.

As with other arms control treaties, provisions exist that specify the duration of the Convention and the inherent right to withdraw from the convention if extraordinary events jeopardize the supreme interests of a nation. The Convention is of unlimited duration. In the case of supreme national interest, a country could withdraw from the Convention with three months notice to the United Nations Security Council. Article XIII provides these necessary assurances.

Article XIV establishes the national requirements for acceding to the Convention, including the time lines for this accession. It simply reflects directions for those wishing to accede on the proper mechanisms for joining the BWC.

Finally, Article XV specifies that the languages of the Convention will be English, Russian, French, Spanish, and Chinese and that texts shall be deposited in the archives of the depositary governments. In true arms control fashion, several of the BWC signatories have provided declarations and reservations. These statements are largely "clarifications" provided by individual states that illuminate certain aspects of the BWC.[26]

CONCLUSIONS

And so with fifteen articles of a total of slightly over 1,700 words, the BWC entered into force in March 26, 1975. In some regards, the beauty in the agreement has been the absolute prohibition against biological weapons, the first arms control treaty to ban an entire class of weapons. However, over time we have come to see the difficulty with such a brief, strategic statement of intent. It has led to uncertainties both in implementation and interpretation that has created a divisive environment that in some respects has led to an international paralysis concerning the BWC. Further, and now importantly with the proliferation of biotechnology, this has contributed to what I believe makes the Convention the "most important arms control treaty of the twenty-first century."

Despite the great accomplishment of the BWC, recognized shortcomings were evident. They fell into three important categories: scope, confidence building and verification, and worldwide application.[27] All of these shortcomings exist in some form today, although some progress has been made across all three areas.

The scope of the BWC did not originally pertain to biological or toxin weapons directed against plants and animals. As noted above, this shortcoming has been addressed. Another issue pertains to the allowance for conducting research on biological agents or especially dangerous pathogens for "prophylactic, protective or other peaceful purposes", thereby providing a potential loophole.

With regard to confidence building and verification, no provisions were developed. Here, minimal progress has been made. Confidence-building measures (CBMs) have been developed, and annual submissions do occur. But the submissions are voluntary and to date participation is paltry, with only about 70 of the 168 member nations participating. What submissions are provided varies widely and demonstrates a significant lack of consistency. Many have also called in question the relevance of the information provided.

At the time of entry into force, no direct verification mechanisms were provided. This has been and remains problematic. Verification is a highly charged issue that detracts from the ability to make progress on other issues. Additionally, at the time of the entry into force, no mechanisms existed for assessing compliance or dealing with violations of the Convention. Largely, these same conditions exist today, although the potential exists for Article VI to serve as a starting point.

Finally, with regard to application, the BWC still lags behind both the CWC and NPT—its chemical and nuclear relations—in the number of member state parties.

All of these issues have become the work program for the review conferences held every five years.

Chapter Five

The BWC Review Conferences (RevCon) and Special Reviews

In March 1975, the BWC entered into force and ushered in a new era in biological and chemical arms control agreement, closing some of the perceived shortcomings of the 1925 Protocol for the Prohibition of the Use in War of Asphyxiating, Poisonous or Other Gases, and of Bacteriological Methods of Warfare. First, the BWC provided an unequivocal norm against the development, production, stockpiling, or acquiring or retaining biological weapons. The Convention also provided the necessary language and thereby the political leverage to move forward on negotiations on the elimination of chemical weapons.

So while there were clearly reasons for optimism that an agreement on biological weapons had been reached, many understood the reality that much work remained to be done on the implementation of the Convention. This work would fall to the review conferences.

The BWC Review Conference (RevCon) process was established by Article XII of the original Convention. The framers of the BWC well understood that a recurring forum for having a dialogue on biological warfare issues was necessary to keep the Convention relevant. The seven BWC RevCons, as they are called in shorthand, occurred in 1980, 1986, 1991, 1996, 2001, 2006, and 2011. Most have been very civilized, with the exception of the 2001 RevCon during which the United States left the negotiations.

It is not my intent to conduct an exhaustive examination of the RevCons or to try to capture each of the "puts and takes" involved in these negotiations. This has been well chronicled, and much of the documentation has been provided for the reader's review within the appendixes. Rather, the desire is to discuss the significant outcomes of the RevCons to capture the progress that has been made as well as to look at the future of the BWC.

The format of the "typical" RevCon is worth a mention. The RevCons are held in Geneva, Switzerland, at the United Nations Mission in the Palais des Nations in a cavernous hall that looks like the inside of the Star Wars Death Star. Nations sit in alphabetical order in concentric semicircles facing a small, elevated stage where the president of the review conference, other conference leaders, and the implementation support staff sit. The conference normally proceeds along a three-week choreographed sequence. One description states,

> The Conference is scheduled to start with a video message from UN Secretary-General Ban Ki-moon, and then to meet in general debate for two days. It is then scheduled to enter into an article-by-article review in the guise of the "Committee of the Whole" (CoW). Towards the end of the middle week of the Conference, the "Drafting Committee" is expected to be convened to translate the work of the Conference into a final report and declaration.[1]

While this description is accurate, it does not fully paint the picture of how the conference proceeds. The first week is posturing. This is the week of the dignitaries; there are lofty speeches highlighting aspirations for a successful outcome. There are also cocktail parties and the renewing of friendships. The BWC community is fairly small, and therefore many of the key national actors are known to the other delegations. At the end of the first week, the national delegations have begun to settle in. The second week begins the substance of the conference. Measuring begins in an effort to see where the key delegations are on the important issues of the day. It is important to note that many of the key issues to be discussed are agreed upon at the preparatory conference (PrepCon) held about six months prior; this meeting establishes the work plan and governance structure for the RevCon. Also in the second week, the short knives begin to come out, positions begin to be developed, and alliances are formed. The second week also begins the article-by-article reviews. By the end of the second week the terrain is fairly well mapped out, and delegations are left to determine what their redlines will be in crafting the RevCon final document. The third week is all about the preparation of the final consensus document. The BWC RevCon ends with the issuing of the final document and congratulations on a job well done.

A thought about this type of negotiation is in order. If 150 or so nations—some BWC member nations do not attend the RevCon for a variety of different reasons—come together for a short, three-week period to negotiate complex issues, developing solutions is highly unlikely. If you further establish that measure of effectiveness is a consensus document at the end of the three weeks, dealing with the most important issues in anything more than a declaratory manner simply does not occur.

Still, the RevCon is the structure that the BWC has established, and it is therefore necessary to work within it to be successful. This is one of the reasons why work done unofficially through pre-RevCon conferences, traveling to capitals to build support for national positions and developing strong alliances prior to the RevCon, becomes critical.

In this discussion of the RevCon process, no attempt will be made to provide an exhaustive review of the sessions. The full, final documents for all review conferences, meetings of state parties, and the intersessional meetings are provided on the United Nations in Geneva website and are available for viewing.[2] Rather, the major outcomes of each of the conferences will be provided. As the First Review Conference both sets the stage for the later conferences and illustrates the degree to which the issues continue to be revisited, additional analysis of this meeting will be provided.

THE EARLY YEARS AND THE FIRST REVIEW CONFERENCE

The early period of entry into force and prior to the first review conference was a time of reflection and coming to terms with what the convention was and what it was not. The reaction of the five permanent members of the United Nations Security Council provides a microcosm of the spectrum of concerns that developed almost immediately. The United Kingdom and Soviet Union declared that neither had offensive BW programs and therefore had no stocks to report as having been destroyed. China and France were unwilling to become parties to the BWC. Neither had participated in the negotiations, and France was concerned about the lack of verification mechanisms. The United States quickly declared that it had had an offensive program and had possessed offensive BW stockpiles but that the program had been dismantled and the stocks destroyed or turned to civilian peaceful use as required by the Convention.

An analogous situation is buying a used car that looks great in the showroom and even during the test drive, but after the first few weeks, the owner realizes that the passenger side window does not work, the car pulls to the left, and there is a slow leak in the right rear tire. You can either live with the annoyances or attempt to correct the deficiencies. The parties to the agreement have generally realized that having the BWC—or in our analogy, the less-than-perfect car—is better than having nothing at all.

Furthermore, the issues that were and continue today to plague the BWC came into clear focus. The early days of the Convention and even during the First RevCon were consumed with questions concerning the implications of science and technology, particularly whether novel biotechnical capabilities would be covered under the convention or render the treaty obsolete. In the end, parties reached consensus on the applicability of the treaty even in the face of ground breaking biotechnical advances. Coinciding with BWC entry into force was the Asilomar Conference on Recombinant DNA. During this meeting, a group of 140 professionals (primarily biologists and other scientists, but also lawyers and physicians) met in February 1975 at the Asilomar State Beach in California to discuss biohazards and the regulation of biotechnology. Their goal was to establish voluntary guidelines to ensure the safety of recombinant DNA technology. The conference also placed scientific research more into the public domain.[3]

The First Review Conference was held in Geneva in March 1980. Major discussion topics included scientific and technological developments, proposals for improving compliance mechanisms, and the exchange of information. The lack of a compliance or verification protocol served as an irritant to a large and growing number of member states. Several nations attempted to finesse the issue by developing techniques that would strengthen compliance and provide more concrete measures for lodging complaints rather than the very general language contained in Article VI. Sweden proposed a consultative committee that would function between the states and the United Nations. This body would have had responsibility for doing fact-finding, investigations, and even on-site inspections to resolve claims of noncompliance. This would have allowed for complaints to be examined by a lower-level body prior to being referred to the UNSC.

Proposals for strengthening compliance generally centered on Articles V and VI. This approach was generally favored by the Western nations. The United Kingdom had a proposal to add language to Article V. Others favored a traditional approach to arms control verification, with a complete verification regime including data exchanges, on-site verification, and challenge inspections.

As a way of increasing the rigor of implementing the BWC, the final document called for "voluntary declarations" to be made concerning previous activities related to offensive BW programs. The language in the final declaration of the First Review Conference served as a foundation for the confidence-building data exchanges that would be adopted in the Second Review Conference.

In a signal of things to come, on the final day of the review conference, the United States declared that discussions were ongoing with the Soviet Union to address concerns about an anthrax outbreak in the town of Sverdlovsk.

The final document established a pattern that would continue with all subsequent review conferences. After some three weeks of discussion, the results came down to a five-page article-by-article review that largely recognizes, declares, and affirms the BWC, making little in the way of substantive change to the original convention. Of course, the full review conference document is much longer—178 pages—and provides a more granular look at the issues of the conference.

During the deliberations, the conference attendees noted with some satisfaction that "no complaints had been lodged" and therefore concluded that Articles I–IV had been successfully implemented.[4] It is curious that the United States refrained from lodging a formal complaint about the Sverdlovsk anthrax incident that had occurred a year earlier. Discussions were also held calling for the sharing of experiences on national legislation. This, too, has been a recurrent and important theme in subsequent review conferences.

Another theme that emerged was the developing nations' complaint surrounding the application of Article X. This theme became part of the final conference language. Specifically, developing nations see this article as calling for promoting "economic and social development" resulting from the disarmament process. The language calls for the "transfer and exchange of information, training of personnel and transfer of materials and equipment on a more systemic and long-term basis."[5]

Still, the unease of the member nations with the Convention's lack of a compliance or verification mechanism was clearly evident. This theme continues today.

THE SECOND REVIEW CONFERENCE

The period between the first and second review conferences was an interesting test for the Convention. Issues surfaced about the Soviet program, particularly concerning two incidents. The first was the Sverdlovsk anthrax incident, which at the time was reported to have been caused by contaminated meat. At first the explanation seemed possible, given that anthrax is a naturally occurring disease endemic to the region. However, over time as word of the incident spread, it became clear that the symptoms and infected area were inconsistent with a naturally occurring outbreak of the gastrointestinal form of the disease. Almost fifteen years later, the actual cause of the infections were confirmed to be the accidental release of weaponized anthrax spores resulting from negligence of one of the plant workers. The worker had failed to install an end-stage filter that would have prevented such a release. While the mortality and morbidity numbers vary widely, what is clear is that at least sixty-six people died and several thousand were exposed and received antibiotic treatment.

The second incident involved allegations of Soviet use of biological or chemical weapons in Southeast Asia in 1975.[6] During this period, reports surfaced of "yellow rain" causing death and illness in several countries in the region, including Vietnam, Cambodia, and Laos. Similar allegations of the use of toxic substances were made against the Soviets in Afghanistan. The evidence was largely circumstantial and certainly not definitive as the affected nations prohibited on-site investigations from being conducted. The

collected samples were tested and confirmed to contain trichothecene myco-
toxins, which were not indigenous to the affected countries, but still it was
not possible to definitely establish a causal link between the Soviet Union
and these "yellow rain" incidents.[7]

The collective effect was chilling. In the Sverdlovsk incident, Russia
eventually admitted responsibility for the release. However, the "yellow
rain" incidents were never confirmed either to be deliberate or of Soviet
origin. Both clearly highlight the difficulties associated with conducting forensics and attribution on biological incidents. Even given overwhelming
circumstantial evidence, establishing definitively that an attack has occurred
remains a matter of judgment. One reference highlights that confidence in the
Convention was seriously damaged by these two incidents, especially in the
United States.[8] This helps to explain why the United States continues today
to have concerns with such an imperfect type of compliance or verification
mechanism that would be more about subjective judgments than factual information. In an odd twist and related to the experiences with the Sverdlovsk
and yellow rain incidents, the United States, which had originally not opposed a verification regime, began to oppose such a mechanism, while the
Soviet Union and the Warsaw Pact uncharacteristically began to favor a
verification mechanism.

It is also noteworthy that during this period, negotiations were ongoing by
an Ad Hoc Group considering the parameters of what would become the
Chemical Weapons Convention. In yet another irony, the emerging language
from the Ad Hoc Group stated the importance of a verification protocol,
including it as one of three topic areas: scope, verification, and other matters.[9]

Another result of the controversial incidents was the 1982 call by several
Western and nonaligned nations for the United Nations to establish procedures to investigate reports of chemical and biological weapons. This request
to the secretary general demonstrated the perceived inability of the BWC to
deal with alleged instances of noncompliance.

The Second Review Conference was held in Geneva in September 1986.
The meeting began with the United States accusing the Soviet Union of
continuing to maintain an offensive biological weapons program. The Soviet
Union denied the allegations. Despite this initial engagement, neither side
pressed the issue during the remainder of the review conference. It was also
during this review conference that the U.S. position with regard to verification became clear and unequivocal. In short, the United States did not support legally binding provisions for strengthening consultation and verification. The final document from the review conference confirms one hundred
member state parties in attendance and notes affirmatively that all permanent
members of the United Nations Security Council are now parties to the
agreement.

During the Second Review Conference, the impact of scientific discovery and biotechnology was considered as a major topic for discussion. The United States and Soviet Union each presented papers on the issue. The U.S. paper expressed concern about genetically modified organisms and concluded that advances in the biotechnology field were making capabilities more readily available, and these changes would further complicate development of any sort of verification mechanism. The Soviet paper considered two new types of pathogens, prions and human immunodeficiency virus (HIV, the virus that causes acquired immune deficiency syndrome [AIDS]), and they reached similar conclusions about advances in biotechnology. The conclusion of the Soviet paper was that the Convention continued to have relevance for natural and synthetic microorganisms and toxins.[10] These papers formed the basis for the Article I assessment that the Convention continued to have broad relevance:

> The Conference reaffirms that the Convention unequivocally applies to all natural or artificially created microbial or other biological agents or toxins whatever their origin or method of production. Consequently, toxins (both proteinaceous and non-proteinaceous) of a microbial, animal, vegetable nature and their synthetically produced analogues are covered.[11]

It is ironic given what we now know of the timing and technical aspects of their program that the Soviet paper reached these conclusions. During this time frame, the program had grown to over sixty thousand people and entailed a wide variety of experimentation on naturally occurring and genetically modified pathogens, thus the paper affirmed that their activities were in violation of the BWC. Of course, this would not be confirmed until over a decade later.

The final document contained an important section on national implementation, noting shortfalls existed by many member state parties and calling for additional emphasis on the development of measures designed to guarantee compliance within national territory, physical protection of laboratories, and inclusion of these issues in textbooks and in medical, scientific, and military educational programs.

Perhaps the most significant and long lasting of the measures undertaken was the establishment of voluntary, annual data exchanges. These exchanges were considered under the Article V discussion, and while not directly linked to verification, they would be an inherent part of any verification mechanism. Information specified for sharing included details on biological laboratories, outbreaks of infectious diseases, publication of results of biological research related to the Convention, and promotion of scientific and technical contacts for peaceful purposes. The intent of the exchange of this information was to decrease secrecy and to promote confidence and cooperation in permitted biological activities. These annual exchanges were politically, not legally, binding, meaning that they were voluntary and not required as part of a state's obligations to the BWC. Essentially, these exchanges became the foundation for the modern Confidence-Building Measures (CBMs) submission. Following the conclusion of the Second Review Conference, a group of scientific and technical experts met in Geneva in March through April 1987 to discuss the modalities of the information exchanges. The CBMs were introduced in the 1986 review conference, expanded in the Third Review Conference in 1991, and are now being considered for refinement following the 2011 Seventh Review Conference.

Article X also received considerable attention in the final document. A listing of recommended measures for increasing cooperation included transfer and exchanges of bioscience information; transfer of materials and equipment; promotion of contacts; greater technical cooperation; facilitating bilateral, regional, and multiregional agreements; and encouraging the coordination of national and regional programs. Of particular note is the call for greater "co-operation in international health and disease control." This last proposal provided a direct linkage to the current growing collaboration of international organizations such as the WHO, FAO, and OIE, to name a few. The final document also notes that no provisions should be developed that would impose restrictions on transfers for peaceful purposes.

Despite the initial concerns voiced by the United States concerning possible Soviet noncompliance, the verification topic, while discussed primarily in the context of Articles V and VI, was not directly considered as a stand-alone issue. This would change during the Third Review Conference.

Topics identified for discussion in the Third Review Conference included the impact of scientific and technological developments; relevance of the results achieved in the chemical weapons negotiations; the effectiveness of Article V and of the cooperative measures agreed to in the final declaration; and whether additional cooperative measures were needed or legally binding improvements, or some combination of both.[12]

THE THIRD REVIEW CONFERENCE

The Third Review Conference was held in Geneva in September 1991. It was the first held in the post–Cold War era and in the immediate aftermath of the invasion of Kuwait by Saddam Hussein's Iraqi army. UNSCR 687 had been approved, which became the basis for the UNSCOM inspections that would follow. With the topics identified in the final document of the previous review conference, the stage was set to discuss the verification/compliance issue as a central focus of the review conference.

The review conference formalized the CBM process began in the previous BWC review, providing the structure for the annual voluntary data exchange measures; this format has remained largely unchanged since 1987. Appendix C provides the agreed-upon format that was developed by the Ad Hoc Group (AHG). The goal of this effort was to improve the poor reporting by member states as well as to provide clarification and the addition of information exchange measures.

The review conference also sought to strengthen the consultative capacity of the BWC, providing procedures for states to raise compliance questions. Contained in the final declaration of Article V is the stated goal of moving toward a proper verification regime:

> The Conference, determined to strengthen the effectiveness and improve the implementation of the Convention and recognizing that effective verification could reinforce the Convention, decides to establish an Ad Hoc Group [AHG] of Governmental Experts open to all States parties to identify and examine potential verification measures from a scientific and technical standpoint. [13]

This AHG (which is different from the previous group) was to be chaired by Ambassador Tibor Toth (Hungary) and to convene from March 30 to April 10, 1992. The group became known as the Verification Experts (VEREX). The language for the group's charter is quite prescriptive, calling for examining potential verification measures in terms of stated criteria, including: [14]

- their strengths and weaknesses based on, but not limited to, the amount and quality of information they provide and fail to provide.
- their ability to differentiate between prohibited and permitted activities.
- their ability to resolve ambiguities about compliance.
- their technology, material, manpower, and equipment requirements.
- their financial, legal, safety, and organizational implications.
- their impact on scientific research, scientific cooperation, industrial development, and other permitted activities, and their implications for confidentiality of commercial proprietary information.

As had been customary from the first two review conferences, strong Article X language was also included in the final declaration. The goals for the developing nations was the same, to ensure that they would not fall farther behind in the global biotechnology race. Article X also included additional provisions such as the establishment of a world data bank to assist in the free flow of information.

Topics identified for discussion in the Fourth Review Conference included the impact of scientific and technological developments; relevance of the results achieved in the chemical weapons negotiations; the effectiveness of confidence-building measures agreed in the final declaration; the report of the AHG of governmental experts on verification; the requirements for the requested resources to assist in effective implementation of the Third Review Conference; and a review to assess whether additional Article V provisions are required.[15]

VERIFICATION EXPERTS (VEREX) GROUP AND FOLLOW-UP AD HOC GROUP

The VEREX Group was established within the final declaration of the 1991 review conference to examine the possibilities for verification. Forty-six nations and the World Health Organization as an observer participated in the VEREX Group over the course of several meetings. The first meeting in March to April 1992 met to develop a list of candidate measures that were to be considered for inclusion in a possible verification regime. Following this initial meeting, other sessions were held to examine and evaluate the measures. A final report was prepared and approved during the fourth and final session conducted in September 1993.

The AHG was charged specifically with identifying measures designed to determine if violations of Article I of the Convention were occurring. That is:

> Whether a State party is developing, producing, stockpiling, acquiring or retaining microbial or other biological agent or toxins, of types and in quantities that have no justification for prophylactic, protective or peaceful purposes;
> Whether a State party is developing, producing, stockpiling, acquiring or retaining weapons, equipment or means of delivery designed to use such agents or toxins for hostile purposes or in armed conflict.[16]

In all, a total of twenty-one measures were identified:[17]

- Surveillance of publications
- Surveillance of legislation
- Data on transfer, transfer requests
- Multilateral information sharing
- Exchange visits
- Declarations
- Surveillance by satellite
- Surveillance by aircraft
- Ground-based surveillance
- Off-site sampling and identification
- Observation
- Off-site auditing
- On-site international arrangements
- On-site interviewing

- On-site visual inspection
- On-site identification of key equipment
- On-site auditing
- On-site sampling and identification
- On-site medical examination
- Continuous monitoring by instruments
- Continuous monitoring by personnel

The potential measures fell into two primary categories, including off-site and on-site actions that could be incorporated into a verification regime. The off-site measures could be subcategorized into information monitoring, data exchange, remote sensing, and inspections, while the on-site measures included exchange visits, inspections, and continuous monitoring. The measures ranged from nonintrusive for measures such as monitoring publications to highly intrusive for on-site medical examinations. For each measure a rapporteur was identified who would be responsible for conducting and presenting the analysis.[18]

It was understood *a priori* that no single measure in isolation could possibly be an effective verification regime. Therefore, a successful regime would likely be comprised of several measures used in combination. The rapporteurs were instructed to consider the technical aspects of each measure to include the definitions of the measure, capabilities, limitations, potential interaction with other measures, and a list of relevant documents. Rapporteurs were also instructed to evaluate their measure singly and in combination. In fact, five representative combinations were provided as examples, such as "declarations, multilateral information sharing, satellite surveillance and visual inspection" and "declarations and information sharing."[19]

The Third Review Conference had directed that a VEREX report would be presented to the states parties for review. This Special Conference, chaired by Ambassador Toth, took place in September 1994, in which eighty states participated. The parties endorsed the VEREX final document, which provided a mandate to "strengthen the effectiveness and improve the implementation of the Convention,"[20] and they further decided to establish another Ad Hoc Group with a mandate to negotiate a legally binding protocol to the BWC protocol.

The follow-up AHG sessions were largely used as forums for the exchange of ideas with nations providing their own papers, yet no final approved document was submitted by the time of the meeting of the Fourth Review Conference.

FOURTH REVIEW CONFERENCE AND THE FOLLOW-UP VEREX
AD HOC GROUP

This review conference was in session from November 25 to December 6, 1996, in between the fifth and sixth meetings of the AHG established through the VEREX Group.

A review of the final document of the Fourth Review Conference provides ample evidence of the lack of progress on any issues of substance. In fact, the final document contained similar pronouncements in many of the articles conducted as part of the article-by-article review as previous RevCon final declarations.

The AHG had made progress on their assigned task of developing a legally binding protocol; however, they had not completed the effort. They did use the BWC review conference to provide an update to the assembled states, but as the legally binding provisions were still under negotiation, little tangible progress could be achieved.

Even the mandate for topics to be considered during the Fifth Review Conference in 2001 left little to be excited about, save the potential for the conclusions of the Special Conference to which the AHG would submit its report.

To this end, the BWC VEREX AHG continued to work during the period from 1997 to 2001. During the seventh session, Ambassador Toth presented what has been called the "rolling text." The text included "158 pages, 22 draft articles, 7 annexes and 4 appendices."[21] By the end of the period of deliberations at the end of the twenty-second session in March 2001, the text had been expanded to thirty articles, yet little agreement could be reached. This growing concern about the lack of consensus led to noteworthy interventions by the chairman. The document states,

> Throughout the two weeks of the twenty-second session, the Chairman conducted a series of bilateral consultations with representatives of States Parties participating in the work of the Ad Hoc Group. The consultations focused on issues in the Rolling Text on which there were strong conceptual differences in views. The consultations were aimed at a conceptual exploration of possible future solutions in the following areas: General Provisions; Definitions; Lists and Criteria, Equipment, and Thresholds; Declarations; Measures to Ensure Submission of Declarations; Follow-up After Submission of Declarations; Consultation, Clarification, and Cooperation; Investigations; Additional Provisions on Declarations, Visits, and Investigations; Confidentiality Provisions; Measures to Redress a Situation and to Ensure Compliance; Assistance and Protection against Bacteriological (Biological) and Toxin Weapons; Scientific and Technological Exchange for Peaceful Purposes and Technical Cooperation; Confidence-Building Measures; The Organization; National Implementation Measures; Legal Issues; Lists and Criteria (Agents and Toxins); List of Equipment; Annex on Investigations; Annex on Confidentiality Provisions.[22]

One account pessimistically summarized the state of the rolling text as having "very little consensus about a protocol less than a year before the Fifth Review Conference."[23] By way of a exclamation point, Ambassador Donald Mahley's July 25, 2001, statement at the AHG provides perhaps the clearest articulation of the United States' stance on the verification protocol. It has been included in its entirety in Appendix F.

The U.S. position was that after more than six years of work, the VEREX AHG draft protocol was not capable of achieving the requirements of the mandate. Simply stated, the United States had "issues with both individual proposals and the general approach to some issues throughout these negotiations." As an example, the AHG rolling text had not been able to find a "mechanism suitable to address the unique biological weapons threat." Rather, the solutions were based in the generalized world of arms control, rather than the very unique circumstances presented by the biological defense issues.

The United States was further concerned that the approach of the AHG would do little to deter those seeking to develop biological weapons, arrive at a framework that would not provide the necessary confidence in compliance, and perhaps even jeopardize our national security and place confidential business information at risk.

While the substance and tone was shocking to some, it reflected a growing unease with the emerging rolling text that the United States had been highlighting during the AHG deliberations. Perhaps the shock was the completeness of the U.S. objections to the ongoing negotiations. Heretofore, the objections had been discussed individually as the text was being prepared. This had been the first time that they had been spelled out in their entirety.

Ambassador Mahley's statement ended on a positive note, highlighting the U.S. desire to work toward more appropriate solutions and pledged U.S. support to "work hard to improve—not lessen—global efforts to counter both the BW threat and the potential impact such weapons could have on civilization."

It is also noteworthy that Iraq was never mentioned during the review conference despite the ongoing inspections and the strong suspicions and eventually the admissions of an offensive BW program in violation of their obligations under the Convention.

FIFTH REVIEW CONFERENCE

The Fifth Review Conference must be placed within the unique context of the moment. The preparatory work had been done in April 2001 prior to the U.S. statement at the twenty-fourth session of the VEREX AHG. Therefore, the agenda and work plan reflected the expectation of having the report to the review conference of the successful text that had been negotiated. Instead, only four months earlier, the United States had directly laid out its strong objection to the emerging draft verification protocol. Additionally, the review conference had begun a couple of months after the terrorist attacks of September 11th and in the immediate aftermath of the anthrax letter attacks. While details of the 9/11 terrorist attacks were emerging and the perpetrators of the attacks were known to be Osama bin Laden and al-Qaeda, little was known about the perpetrators of the Amerithrax attacks, as they have come to be known.

Were the Amerithrax attacks the work of a terrorist or a state? If they were a terrorist attack, were they state sponsored? These answers would take almost eight years to answer, and even today questions persist. Still, this was the context under which the Fifth RevCon officially opened on November 19, 2001.

A cursory look at the final report of the review conference provides little evidence of the contentious nature of the session. The one clue is the lack of an article-by-article review and the brevity of the text. The Fifth Review Conference was the first and only periodic meeting of the member states to have been suspended. Officially, the meeting was held from November 19 to December 7, 2001, and it concluded without a final document and reconvened from November 11 to 22, 2002. The final document reflects little of the turmoil leading up to the conference or the 11 month suspension. In fact, the final document looks similar to the other predecessor documents of previous RevCons.

As is customary, the review conferences begin with senior officials from national capitals arriving at the conference to deliver speeches about the importance of the efforts that will be expended and the need to reach successful outcomes. The U.S. senior leader for the Fifth Review Conference was John Bolton, the Under Secretary of State for Arms Control and International Security. His remarks, like Mahley's during the final AHG meeting, were clear and to the point:

> The United States will simply not enter into agreements that allow rogue states or others to develop and deploy biological weapons. We will continue to reject flawed texts like the draft BWC Protocol recommended to us simply because they are the product of lengthy negotiations or arbitrary deadlines if such texts are not in the best interests of the United States.[24]

As Mahley did previously, Bolton spoke of flawed approaches contained in the "rolling text" and the need to think differently about the biological threats we face. He also named several states of concern, including Iraq, North Korea, Iran, Libya, and Sudan.

During the review conference, attempts were made to salvage the work of the AHG and resolve the differences over "rolling text." However, in the end, the United States proposed that the Conference terminate the AHG's mandate. The United States considered the draft protocol flawed, failing to consider relevant issues of the day, such as biological terrorism.

While the United States was clearly displeased with the work of the AHG, other nations also had serious concerns. The necessary interventions by the chairman, Ambassador Toth, at the twenty-second meeting of the AHG provide indications that their efforts were at best not universally endorsed and at worst seriously flawed. Based on the U.S. objections, the AHG negotiations ended in failure in 2001, and the Fifth RevCon was suspended. A consensus decision was reached to adjourn the Conference until November 11, 2002.

The final document of the Fifth Review Conference contains little of substance. The only major outcome was the agreement to hold three annual meetings prior to the beginning of the Sixth Review Conference "to discuss, and promote common understanding and effective action on":

1. the adoption of necessary national measures to implement the prohibitions set forth in the Convention, including the enactment of penal legislation;
2. national mechanisms to establish and maintain the security and oversight of pathogenic microorganisms and toxins;
3. enhancing international capabilities for responding to, investigating, and mitigating the effects of cases of alleged use of biological or toxin weapons or suspicious outbreaks of disease;
4. strengthening and broadening national and international institutional efforts and existing mechanisms for the surveillance, detection, diagnosis, and combating of infectious diseases affecting humans, animals, and plants;
5. the content, promulgation, and adoption of codes of conduct for scientists. [25]

Given the contentious manner in which the Fifth Review Conference ended, one could reasonably conclude that the future of the BWC was in serious jeopardy. The United States had felt it was necessary to intervene but never intended to place the BWC at risk. In fact, statements from Bolton and Mahley strongly indicated the relevance of the BWC; the issue was one of tactics.

The annual meetings became the foundation for what today has become the intersessional process. The meetings were designed to be informational rather than decision-making forums, and as such they were most important for keeping the dialogue going concerning BWC issues.

SIXTH REVIEW CONFERENCE (NOVEMBER TO DECEMBER 2006)

The Sixth Review Conference was held in November to December 2006. Fresh from the newly minted intersessional process and the three annual meetings, the focus was on national implementation, surveillance for and responding to biological incidents, and codes of conduct for scientists.

Examples of some of the more noteworthy outcomes include adoption of necessary national measures to implement the prohibitions set forth in the Convention, including the enactment of penal legislation; national mechanisms to establish and maintain the security and oversight of pathogenic microorganisms and toxins; enhancing international capabilities for responding to, investigating, and mitigating the effects of cases of alleged use of biological or toxin weapons or suspicious outbreaks of disease; strengthening and broadening national and international institutional efforts and existing mechanisms for the surveillance, detection, diagnosis, and combating infectious diseases affecting humans, animals, and plants; and the content, promulgation, and adoption of codes of conduct for scientists.

Another important inclusion was the reference to the United Nations Security Council Resolution (UNSCR) 1540, which places obligations on all states and "affirms support for the multilateral treaties which aim is to eliminate or prevent proliferation of nuclear, chemical or biological weapons."[26]

With respect to Article X, the language continued to be strengthened, calling for specific collaboration with member nations and other international organizations and networks. These crosscutting efforts were intended to be with the "WHO, FAO, OIE and IPPC" [Intergovernmental Panel on Climate Change] on issues of mutual concern such as epidemiological issues and public/animal/plant health, biosurveillance, and emergency and disaster management plans.

Several important decisions and recommendations were contained within the final document of the review conference. First was the establishment of an Implementation Support Unit (ISU) consisting of three full-time staff responsible for administrative support and several tasks related to the confidence-building measures annual submissions. Interestingly, the final document is quite clear that "the Unit's mandate will be limited to the above mentioned tasks [meaning administrative support and the CBMs],"[27] making it quite clear that there was no stomach for freelancing or perhaps moving toward a verification function.

The final document also reaffirms the intersessional process, providing for meetings on ways and means to enhance national implementation; regional and subregional cooperation on implementation of the Convention; national, regional, and international measures to improve biosafety and biosecurity; oversight, education, awareness raising, and adoption and/or development of codes of ethics; enhancing international cooperation, assistance, and exchange in biological sciences and technology for peaceful purposes; promoting capacity in the fields of disease surveillance, detection, diagnosis, and containment of infectious disease; and provision of assistance and coordination with relevant organizations in the event of the alleged use of biological or toxin weapons. Intersessional meetings were held annually from 2007 to 2010 to discuss these topics. With the creation of the ISU and the development of the intersessional process, the BWC had seen a move toward a new structure for considering the biodefense issues of the day.

The final significant recommendation of the Sixth Review Conference was the call for universalization. The report noted that only 155 nations are states parties and that more should be done to increase membership, noting that "the Convention falls behind other major multilateral arms control, disarmament and non-proliferation treaties."[28]

Perhaps the most important outcomes of the Sixth Review Conference were the conference's reaffirmation of the importance of the BWC, the willingness to continue the biodefense dialogue, and the approach of the international community with respect to the verification issue. By all accounts—and for better or worse—the BWC review conference process had returned to normalcy.

SEVENTH REVIEW CONFERENCE

The Seventh Review Conference was held in Geneva in December 2011. As customary, the results of the previous conference had spelled out the agenda—at least the major topics to be considered—for the meeting. These included new scientific and technological developments relevant to the Convention; progress made on implementation of the Convention; and progress made on the decisions and recommendations from the previous conference.

The context for this review conference was also important. While the collective memory of the participants still remembered the events of the Fifth Review Conference, the successes of the meetings from 2003 to 2010 gave hope for the future. In the United States, a new administration had been elected, and there was hope for a new willingness to collaborate internationally. The Obama Administration had published in 2009 a *National Strategy for Countering Biological Threats* based on Presidential Policy Directive (PPD)-2. The significance of the timing of the document in the life of the administration should not be taken lightly. In typical fashion, PPD-1 established the formation of the structure and process of the White House's national security staff. Countering biological threats was the second national security issue to be addressed by the Obama Administration. In fact, the president's national biological strategy was first unveiled publicly at the 2009 BWC Meeting of States Parties (MSP) by Bolton's successor several

times removed, Under Secretary for Arms Control and Nonproliferation Ellen Tauscher.

The review conference also followed a particularly intense period of
internal interagency deliberation, which saw the development of national
goals and objectives for the 2011 RevCon. These included development of
language for Article X, science and technology, biotech transfer, confidence-
building measures, universality, national implementation, and verification
and enforcement mechanisms. In developing these programs of work, the
interagency stayed far away from the issue of the return to a verification
protocol.

International conferences were held in advance of the actual review conference to provide opportunities for dialogue and for gaining an understanding of national positions. They included Wilton Park (2010), Beijing (2010),
Montreux (2011), Berlin (2011), Manila (2011), and Clingendael (2011).[29]
These conferences were notable from the standpoint of seeming to be very
reasonable, with attendees attempting to find issues of common ground
where progress could be made without getting into the still-contentious verification protocol.

Unfortunately, despite all of the promise, the Seventh Review Conference
accomplished little and most notably continued what remains a very unpro-
ductive verification protocol debate. One review of the RevCon begins,

> The December 2011 review conference of the Biological Weapons Convention
> (BWC) demonstrated the danger of the bioweapons ban drifting into irrele-
> vance. Standstill was the motto of the meeting. Only incremental improve-
> ments on some procedural issues were achieved.[30]

This is certainly not a rousing endorsement of the work that was done at the
meeting. Several perceived missed opportunities have contributed to the malaise surrounding the 2011 meeting. No progress was made for dealing with
advances in biotechnology. Despite the breakneck pace of developments in
this Age of Biotechnology, no new ideas or mechanisms for assessing the
impact on the BWC were developed. Even gaining a collective and shared
understanding of the threats we face has been problematic. Some member
nations want to keep the treaty in its pure form as an arms control forum,
while others would prefer to see a more expansive view of the applicability
of the Convention to include relevant biodefense issues of the day. Furthermore, despite the Obama Administration's and even several international
partners' hopes for avoiding the verification issue, the final days of the
Seventh Review Conference will be largely remembered by an end run by
the group that became known as the PRIICs, for Pakistan, Russia, India, Iran,
and China.

The PRIICs' other agenda may well be "to recover and assert ownership of the BTWC by the States Parties" in contrast to the move toward opening the aperture with respect to the applicability of the Convention.[31] This debate has continued to smolder. Does the BWC only pertain to state-to-state biodefense issues? Or is it relevant to the bioterrorist threat? And what about its applicability to naturally occurring disease and organizations such as the WHO, FAO, and OIE?

The resurfacing of the verification debate also caused some consternation and opened old wounds as several European states actually favored a verification protocol as a long-term outcome, yet were willing to take a more measured approach until the United States was willing to come along. Additionally, it is not clear how much there is a genuine call by the PRIICs for a verification regime for the BWC and how much is related to needling the United States on this issue. Several of these states are undoubtedly doing so for theater and political gamesmanship.

The participants demonstrated no willingness to have a BWC with greater structural and organizational capacity to handle the biodefense issues of the day. The recommendation to increase the size of the ISU was not approved. The measure failed as a result of the potentially modest increases in national assessments that in some cases would have been measured in the tens of dollars. The proposal to have limited decision-making authority outside of the RevCon was likewise disapproved; essentially, this dictates that decisions will continue to be made every five years.

In all, the final document lists 103 member nations that participated along with five signatories that have not ratified, two observer nations, eight non-state observers, and forty-seven nongovernmental organizations that attended the conference. The article-by-article review was short and at best *pro forma*. The intersessional process had proven valuable, and agreement was reached that the meetings would be held during the period from 2012 to 2015. Even the topics to be discussed at the 2016 review conference were uninspiring:

1. new scientific and technological developments relevant to the Convention, taking into account the relevant decision of this Conference regarding the review of developments in the field of science and technology related to the Convention;
2. the progress made by states parties on the implementation of the Convention;
3. progress of the implementation of decisions and recommendations agreed upon at the Seventh Review Conference, taking into account, as appropriate, decisions and recommendations reached at previous review conferences.

At the start of the Review Conference, the president of the meeting, Paul van den Ijssel from the Netherlands, had declared the mantra would be "ambitious realism." This is what he got, although likely not in the way that he had envisioned. The first days of the conference were clearly ambitious, with the usual upbeat declarations of hope. A highlight was the arrival of Secretary Hillary Clinton, the U.S. Secretary of State, who delivered the national address to the assembled delegates. The second part of the conference had been the realism. In a daily RevCon report prepared by the BioWeapons Prevention Project (BWPP), the author writes,

> This was a difficult Review Conference in its later stages, much harder than that in 2006. The contrast between the positive aspiration tone of the first week and the tensions of the final few days was stark. Many delegates were painfully aware that they weren't able to achieve all they wanted. There were some minor successes—the new Article X database has potential and the appointment of two Vice-Chairs each year of the ISP [Intersessional Process] will provide additional political focus—but there were greater losses.[32]

The author goes on to lament the inability to make progress on the verification issues, describing the United States as having "a deep-seated fearful reaction to the verification 'bogeyman.'" This translated into an inability to even agree to discuss verification issues during the intersessional meetings.

Curiously, a major topic of the RevCon turned out to be the global economic crisis. This translated to a failure to approve the ISU going from three to five full-time personnel due to the fiscal implications of such an increase. BWPP RevCon report #14 reports that such an increase would have meant that "26 States Parties would pay something like only US$19 per year in [additional] contributions," certainly affordable. Yet some nations had been given the mandate for zero growth in national assessments and therefore could not approve such a proposal.[33]

To put this into perspective, in a meeting held every five years, a surprisingly significant amount of time was spent on budgetary issues and not on the relevant biodefense issues of the day. On balance, it is hard to envision much less being accomplished during a review conference. The final lines of the BWPP RevCon report from the Seventh Review Conference sums it up nicely:

> The regime to control biological weapons, of which the BWC is the focal point, is built upon the convergence of legal, political, scientific, technical, moral and humanitarian (including public health) issues. A major weakness of the Seventh Review Conference was the focus of some delegations on a purely legalistic perspective rather than a focus on practical action that could reduce biological threats around the world. When histories are written of the BWC in decades to come, 2011 will be seen as a significant missed opportunity.[34]

CONCLUSIONS ON THE REVCON PROCESS

Let me begin by stating that I am a pragmatist at heart. In fact, I am willing to negotiate at the drop of a hat and fully understand that in a negotiation, each party likely fails to achieve the full measure of its initial goals and objectives. This is called compromise. Unfortunately, the BWC RevCon process has shown a significant intransigence on the part of several key member states parties and almost a complete inability or unwillingness to compromise.

My own introduction to the emotion of the verification issue occurred at an international conference where I served on the U.S. delegation. In a private conversation outside of earshot of any international delegates, I mentioned the *V*-word, *verification*, to an interagency colleague and received a fifteen-minute tongue-lashing about how in the United States we do not even mention verification, we only talk about compliance. I had clearly hit a raw nerve. This sensitivity continues today, with few even willing to entertain a discussion of the topic.

At the time of the signing of the BWC, many compromises were made in order to rapidly achieve a politically acceptable document. These compromises continue to plague the RevCon process. This became the expedient way to achieve an important political outcome, the elimination of biological weapons, at the expense of taking more time and effort to develop a verification or, if you like, a compliance mechanism that would enhance confidence that members were adhering with the terms of the Convention.

Today, the RevCon lacks the ability to manage the velocity of the BW issue. Progress resulting from the RevCons is measured in glacial terms for a science moving at the speed of light, causing an impedance mismatch. This mismatch may ultimately lead to the growing irrelevance of the BWC as anything more than a statement of political will. Maybe that will be enough as common practice has been to consider WMD, including BW, as a low-probability, high-consequence event. On the other hand, if one believes that the BWC has relevance beyond state BW, then such obsolescence becomes problematic.

On the tactics of the review conferences, the article-by-article reviews are mind-numbing and accomplish very little. Even the language of the reviews from RevCon to RevCon has remained largely the same, with only minor tweaks. The structure of the three-week conference hinders significant progress and certainly does not contribute to the global biodefense dialogue on a real-time basis.

The current BWC methods even call into question the degree to which the international community is serious about the issue. The exclamation point for this comes from the Seventh Review Conference, where nations could not even agree to a small increase of US$19 assessment for the ISU despite the collective agreement about the importance of the organization and the overall positive manner in which it has conducted itself.

Ultimately, several questions must be confronted in determining the fate of the BWC. What are the threats we are facing in this Age of Biotechnology? How best can and should they be addressed? Does the current meeting structure meet the needs of the global biodefense community? How should the BWC fit into mitigating future biothreats? Is verification a necessary part of ensuring future security or simply an unnecessary anachronism from the world of traditional arms control methods that do not apply to biotechnology?

Chapter Six

To Verify or Not to Verify

The central issue facing the BWC is the question of compliance and verification. The Seventh Review Conference has made clear that the future trajectory of the BWC is inextricably linked to future decisions with regard to this issue. Yet at the very time that the trajectory is being set, the BWC debate was somewhat muted, and the final documents were largely silent on the issue. The result is that little has been done to advance the dialogue.

The identification by the BWPP in its final Seventh Review Conference report that the BWC had adopted a "purely legalistic perspective rather than a focus on practical action that could reduce biological threats around the world" allowed the opportunity for revitalizing the Convention to slip away.

This chapter examines the verification question through several lenses to better gain a perspective on the issue. First, scenarios were considered in trying to understand the inherent difficulties in developing a verification regime and the specific difficulties surrounding today's dual-use biodefense issues. Second, the history of verification is considered as a framework for increasing understanding of this complex issue. Finally, specific historical case studies are considered that provide invaluable insights into the verification issue.

UNDERSTANDING THE LIMITATIONS OF THE BWC

I have always found that scenarios can be useful in taking abstract issues and placing them into a context under which they can be assessed and better understood. For this reason, this chapter on verification begins with an examination of scenarios that consider scale, types of activity, relevance, and technical aspects of the Convention. Using these four categories, one can begin to understand what the BWC really limits and perhaps even the parameters under which the Convention can and should function.

The centerpiece for evaluating these scenarios is Article I of the Convention, where each member nation "undertakes never in any circumstances to develop, produce, stockpile or otherwise acquire or retain microbial or other biological agents, or toxins whatever their origin or method of production, of types and in quantities that have no justification for prophylactic, protective or other peaceful purposes."

First is scale. Several years ago, the United States set up a program to examine the feasibility of a terrorist cell developing biological weapons. The program was called Project Bacchus and resulted in a better understanding of how a clandestine bioterror facility could be developed, the results that could be achieved for modest investments, and thoughts on how to thwart such activity. So in asking whether this effort was permissible under the BWC, one would have to reasonably conclude that it was allowable for the government of the United States to conduct this experiment for "prophylactic, protective or other peaceful purposes." The effort was government sponsored, the United States has laws, policies, and regulations for dealing with Biological Select Agents and Toxins (BSAT), and the biological material used was a simulant and not actual pathogenic material.

How about a scenario in which an individual with no relationship to his government is conducting experiments with highly pathogenic material? If the material is not controlled by his national or local laws, then the activity would not necessarily be covered under the BWC. The experimentation may be "allowable," but if the material was deployed in an attack, the person could still be prosecuted under general criminal laws of the nation that was attacked. So in this case, the individual could actually "develop, produce, stockpile or otherwise acquire or retain" a BW capability without necessarily violating the BWC.

Now let's look at a nation-state example. Let's assume that national intelligence finds a suspect facility that includes all the equipment necessary for developing biological weapons. Does this violate Article I of the BWC? Well the answer is, "That depends." What is the intent of the facility? If it was set up as a training facility to train military forces on how to assess and eliminate an adversary's BW site, then it is for protective purposes and therefore allowable. What if this same facility had a plant for the filling of munitions? This seems as if it would be a violation, but since the Convention is an intent-based, rather than capability-based, treaty, a subjective evaluation would need to be made.

Let's now say that the national intelligence finds a suspect facility that includes all equipment necessary for developing biological weapons. Does this violate the confidence-building measures submission if it is not declared? Well again, this is very ambiguous. The CBMs are voluntary and therefore the contents of a national submission are voluntary. However, it clearly calls into question the intent of the member state that had failed to report the facility. By doing so, the nation could have declared that the facility was designed for a specific prophylactic, protective, or other peaceful purpose. Then, this would have been allowed under BWC. On the other hand, nonreporting on the CBMs raises significant questions and may not pass the test in the court of public opinion or the Washington Post test, but still could be legal depending on the intent of the activity.

A state has some secret facilities that it does not wish to report. Is that allowable under the BWC CBMs? Well again, that depends. The BWC makes no provisions for secret facilities, and the CBM submission is voluntary. So technically, the default comes back to what the intent of the facility is. If it is for prophylactic, protective, or other peaceful purposes, then it would be allowed, while if it could be demonstrated that the facility was for offensive BW weapons development, it would be considered a violation of the BWC.

Let's now consider the type of activity. If a nation is conducting an experiment that pertains to one of the NSABB seven experiments of concern, such as rendering a vaccine ineffective, increasing the transmissibility of a pathogen, or allowing the evasion of diagnostic/detection tools, it could be a violation of the BWC, but the test would come down to the intent of the activity. Merely conducting the experiments does not violate the BWC. In fact, the NSABB criteria are not a BWC-approved "test" for assessing violations. Additionally, if the tests were being conducted to understand the limits of the effectiveness of a vaccine, perhaps it could be justified.

If on the other hand, the activity would enable the weaponization of a biological agent or toxin, one would certainly be tempted to conclude that it should be captured under the BWC. Unfortunately, this may not be the proper conclusion. The activity could have been used to develop small aerosol-sized particles to test the effectiveness of antibiotics or antivirals in treating a respiratory infection caused by BW and, therefore, would be permitted under the Convention. Even the use of the term *weaponization* can be problematic. The same dual-use technologies that can be used to deliver aerosol drugs and therapeutics such as insulin can be used equally effectively to deliver BW weapons.

The third set of scenarios concerns the continued relevance of the BWC. The recent dialogue about the suitability of publishing the H5N1 avian influenza articles was ongoing at the time of the Seventh Review Conference, yet it attracted little attention during the meeting and certainly was not included in the final declaration. So here is one of the most important biodefense issues of the day that many believed had the potential to provide a recipe for a bioterrorist to increase the virulence and transmissibility of a seasonal influenza serotype, and it does not become part of the dialogue.

Should it have been a major topic, or would it have disturbed the agreed-upon work plan?

If one were to consider a global pandemic, either naturally occurring or man-made, what would be the mechanisms available for the BWC to take an active role in the response and recovery? Given that the member states parties only convene on an annual basis with the new intersessional process and on a five-year timetable for taking decisions, what relevance does the Convention have for more immediate biodefense issues?

The final group of scenarios concerns the technical aspects of the Convention. Consider definitions. Definitions are essential to arms control agreements, and they certainly are for developing verification mechanisms. They serve to delineate what is in and what is out of the treaty. For example, looking at the Conventional Forces in Europe (CFE) treaty that limited holdings of conventional weapons, the definition of a tank was a self-propelled armored vehicle with a main gun of ninety millimeters or greater. If a vehicle did not meet the criteria, it was not considered a tank. Therefore, the task of identifying and ultimately counting the number of systems becomes a straightforward endeavor, and assessments can be made about a state's level of compliance with the terms of a treaty.

Of course, the danger in relying on definitions is that as a term is defined, one can find ways to circumvent the treaty. Consider the question of "What is a pathogen"? It seems fairly clear, although in application, it is much more difficult to discern. Consider Interleukin-4 (IL-4), which is a bioregulator or short-chain peptide and a naturally occurring component of the human immune system. In large quantities, IL-4 can cause the human immune system to overreact and create a cytokine storm that could prove deadly. Is it a pathogen?

With advances in the life sciences, the very essence of life is being challenged. We are learning about the biological composition of living organisms and developing capacities that will fundamentally alter the way we think about daily life. Longevity is likely to be extended, perhaps even doubled over the next twenty years. Diseases that were thought to be incurable will either be eliminated or made chronic. Life-forms are being altered through genetic engineering. We are seeing life at levels that heretofore have been opaque and understanding that small proteins and changes to gene sequences have dramatic effects.

Can the BWC with its five-year meeting structure and three-person support cell really hope to ensure the premier biowarfare treaty remains relevant and viable? Or should the BWC be looked at as a biodefense vice biowarfare treaty? And can it support a robust verification structure?

WHAT IS VERIFICATION?

Verification is the process that one country uses to assess whether another country is complying with an arms control agreement. To verify compliance, a country must determine whether the forces or activities of another country are within the bounds established by the limits and obligations in the agreement. A verifiable treaty contains an interlocking web of constraints and provisions designed to deter cheating, to make cheating more complicated and more expensive, or to make its detection more timely. In the past, the United States has deemed treaties to be effectively verifiable if it has confidence that it can detect militarily significant violations in time to respond and offset any threat that the violation may create for the United States. [1]
—Amy Woolf, *Monitoring and Verification in Arms Control,*
Congressional Research Service (CRS) Report
December 23, 2011

Verification is a complex activity based on a mix of objective and subjective evaluations that collectively provide a basis for assessing a party's level of compliance with an international agreement. While verification is often considered in absolute terms, in reality, there is far less clarity in compliance assessments. In truth, no treaty can be considered to be completely verifiable. While the result of a nation's assessment of another party's compliance might be the definitive statement such as, "We find Country X to be in compliance," in reality it is more about judgments made based on the assembled facts. A single tank unaccounted for during an inventory may be a technical violation, but should it result in a finding of noncompliance? How about a technical error on the inventory of treaty-limited equipment (TLE)? On the other hand, failure to report a hundred assembled nuclear warheads would likely be both a militarily significant violation and noncompliance issue.

In theory, an arms control verification regime consists of three components: analysis process, evaluation, and resolution. The analysis process consists of the monitoring tools and mechanisms that allow for developing a situational understanding of the facts. Are weapons holdings appropriately accounted for in military facilities? Is there any behavior ongoing or previously conducted that deviates from the established limits of the agreement? What restrictions, obligations, and cooperative measures are contained in the agreement? The evaluation process entails making judgments about the degree to which a party is in compliance. It results from the collective assessment of the level of compliance and not simply any one single piece of the analysis. This evaluation is largely a qualitative, some might even say political, rather than a strictly technical assessment. Finally, the resolution phase of arms control verification is one in which the country assessing the noncompliance to an agreement seeks to resolve the discrepancies. This can be accomplished through a variety of means from political discourse to the use of force.[2]

Verification in arms control began in the late 1950s and largely was conducted in the spirit of the Cold War. The lack of trust between the Soviet Union and United States resulted in arms control agreements that relied most heavily on each nation's national technical means (NTM) to detect violations of treaties, with limited on-site monitoring and little focus on monitoring behavior or exchange of information pertaining to military activities. Arms control verification was focused on having the capacity for detecting militarily significant violations that could undermine the security of a treaty participant if the violation went undetected.

The 1970s, with the Strategic Arms Limitation Talks (SALT), saw the coin of the realm in arms control shift to be agreements that were "adequately verifiable." This nuance allowed that participants would have high confidence of being able to detect significant violations and respond to any security risk. With the Intermediate Range Nuclear Forces (INF) agreement, a shift toward "effective verification" was made that would require being able to detect only militarily significant threats, as well as other violations or discrepancies that might be used for justification for a political response.

Paul Nitze, former Deputy Secretary of Defense, a chief architect of U.S. national security policy, provided what has become a standard definition of "effective verification" while testifying before the Senate Foreign Relations Committee (SFRC) in support of the INF Treaty in 1988:

> What do we mean by effective verification? We mean that we want to be sure that, if the other side moves beyond the limits of the Treaty in any militarily significant way, we would be able to detect such a violation in time to respond effectively and thereby deny the other side the benefit of the violation.[3]

In a sense, Nitze's statement reflects the understanding that arms control is not a perfect science made up of absolute redlines, but rather interpretations that must be tempered by measuring the degree to which a party is meeting its obligations and less about the exact technical findings. It also reflects that at times, different interpretations of the obligations may be held by the parties or some obligations might be difficult to implement, and therefore, it is perhaps not possible to be strictly implemented. Nitze's statement does not constitute a "legal" definition of "effective verification"—in fact, no such criteria have been established. Despite this shortfall, Nitze's statement has gained wide currency (and has survived to be cited with approval by several successive administrations).

Nitze saw critical factors for assessing a verification regime, including: How likely is it that the other side would violate the treaty (and one should ask this for each separate provision in the treaty)? How militarily significant would various types of violations be? How likely is it that we would detect the violations (factoring in not only the treaty's verification mechanisms but also all the other devices for gathering information about other countries)? How quickly would we be able to detect a violation? What sorts of responses, on what timetable, and at what cost, would we be able to undertake to offset the violations and to deprive the cheater of the anticipated benefits (noting that suitable responses need not "match" the violation precisely, in order to be able to offset it)?

THE TOOLS OF VERIFICATION

A full discussion of the tools of verification is well beyond the scope of this book; however, the concepts that undergird a verification regime are necessary to understand as well as the costs and benefits associated with each. Generalizations will be made concerning cost of verification tools; however, a note of caution is in order. The actual cost of implementing a verification regime will be directly related to the degree to which new, unique capabilities must be introduced and the number of unique requirements for verification.

Previously, we discussed the twenty-one measures identified by VEREX that fell into two categories, on-site and off-site measures, with regard to the BWC draft verification protocol. Just as with these deliberations that identified discrete potential combinations of activities for arms control verification, combinations of analytical processes can be considered for incorporation into a verification mechanism.

For our discussions, five categories of tools will be considered. Cooperative measures include transparency and openness activities such as exchanging information on military doctrine, noninterference protocols, and reporting on exercises and other military activities. These measures can also include transparency and confidence-building measures, information exchanges on plans, programs and changes in forces, technical specifications on weapons, demonstrations, and requests to move items into view of national technical means of verification. Cooperative measures also include notification of the movement of treaty-limited items to and between facilities, including bases where the item is stored or deployed and notifications of launch. START I had a large number of legally required notifications. National technical means (NTM) would include capacities such as seismic stations and observation from radar, optical, or satellite surveillance from outside the territorial borders of the inspected nation. Technical monitoring devices, such as perimeter portal continuous monitoring (PPCM), can also be placed at or near sites with treaty-limited equipment. PPCM was used in INF and START. On-site inspections include visits to sites with equipment and storage of weapons or viewing of exercises and exchange visits. Routine inspections are used to monitor activities at declared sites, and per the agreement, inspectors go to a set number of facilities. There can be limits on the number of such inspections per year as in START I. Additionally, it is not uncommon for on-site inspection to include challenge inspections to monitor undeclared or suspected clandestine activities. Such is the case with the CWC, which incorporates challenge inspections by the OPCW and the NPT and which allows for challenge inspections conducted by the IAEA under the provisions of the additional protocol. Finally, sources of information through intelligence channels or open source literature that allows for gaining a better understanding of the weapons holdings and activities of a nation can be used.[4]

At the lower end of the cost spectrum would be nonintrusive measures, such as examining the open literature that requires no cooperative support, while at the higher end would be those site-specific verification measures such as PPCM and on-site inspections. Naturally, the cost of implementation is directly related to the number of interactions required by the monitoring regime (e.g., number of site visits). These costs are not inconsequential for either the host nation or the inspectors. Along with financial costs, it is worth also noting that there is a perceived cost of measures relating to the degree of intrusiveness. The more intrusive a measure is perceived to be, the greater the concern about the potential for intelligence gathering and protecting sensitive or proprietary information.

However, detecting noncompliance is not sufficient. Verification must be accompanied by a means to address compliance issues and to provide for real consequences if obligations not followed undermine confidence in the treaty. To this end, joint consultative bodies provide a means to bring forward complaints and to address compliance issues. Examples for bilateral agreements include the special verification commission (SVC) for the INF Treaty, the Standing Consultative Commission (SCC) for the ABM Treaty, the Joint Compliance and Inspection Commission (JCIC) for START, and the Bilateral Consultative Commission (BCC) for "New START." For multilateral agreements, the IAEA and OPCW for the NPT and CWC, respectively, fulfill this role of providing a forum for the discussions of topical issues and areas of concern.

THE DELICATE BALANCE: DIFFERENT VERIFICATION FOR DIFFERENT TREATIES

A tendency exists to do arms control by analogy, and nowhere is this more acute than in the development of verification protocols. In arms control by analogy, one attempts to relate previous experiences in looking for solutions.

Because this technique worked for this Treaty "X," it should be incorporated into treaty "Y." Unfortunately, much of our collective experience is with nuclear arms control where the focus is rightly on the control of nuclear material as a matter of the highest priority. For nuclear weapons, if one does not have fissile material, then all the know-how in the world will not allow for the development of a nuclear weapon. On the other hand, for biological weapons, the combination of the material that is naturally occurring and the increasingly available capability to process, weaponize, and deploy the pathogen combines to make entering into this field much less difficult.

Looking at the different treaties provides interesting insights. For the nuclear weapons treaties such as START, INF, and New START, more measures have been incorporated into the verification regime. The concern with the loss of control of even a single weapon drives verification tools such as PPCM and on-site inspections. NTM are important, as are cooperative measures such as on-site inspections, transparency and confidence-building measures, and notifications. Continuous destruction monitoring allows for verifying that weapons have been destroyed as agreed.

For the Conventional Forces in Europe (CFE) Treaty, there is less concern about individual pieces of conventional military equipment such as tanks and artillery. Therefore PPCM would not be a useful (or a cost-effective verification tool). In fact, of considerably more interest are the confidence-building measures that allow for observing military exercises. By doing so, the observers can gain an understanding of the adherence to the treaty. Counting individual tanks is less important than counting tank battalions. Understanding if troops are exercising offensive or defensive operations assists in understanding the intent of the forces.

For the CWC, a mix of on-site inspections, transparency, and confidence-building measures and destruction monitoring is used. Once all chemical stocks are eliminated, the destruction-monitoring component of the verification regime will cease. This verification is conducted under the auspices of the OPCW and provides for inspections of military and civilian sites. These measures are augmented with national implementation procedures dictated by laws, regulations, and policies of the armed forces and the private chemical industry.

For the BWC, only transparency and confidence-building measure exchange-of-information has been developed. Additionally, these are voluntary measures and highly open to interpretation. For example, the measures request information on the highest containment level facilities. Yet this phrase is subjective. As with the CWC, these BWC compliance measures have been augmented with national implementation procedures dictated by laws, regulations, and policies of the armed forces and private industry. In addition, during the last review conference in December 2011, Secretary Clinton announced unilateral transparency measures that would include site visits to a U.S. biodefense facility.

More on the transparency visit is in order. The visit was held in July 2012 and by all accounts was well received by the ten international ambassadors who attended the event. For some, the visit likely begs the question of what the U.S. intent was for opening up a biodefense facility in this manner and why then is the United States against a verification protocol that on the surface looks quite similar to the visit. The subtle, yet critical, difference is that a verification protocol would allow for others to assess compliance, while the transparency visit is intended to allow the participants to gain confidence in U.S. compliance.

The Delicate Balance

Nuclear (START, INF, NST)	Conventional (CFE)	Chemical (CWC)	Biological (BWC)
❑ National Technical Means (NTM)	❑ National Technical Means (NTM)	❑ National Technical Means (NTM)	❑ National Technical Means (NTM)
❑ Portal Perimeter Continuous Monitoring (PPCM)	❑ Onsite Inspections	❑ Onsite Inspections	❑ Transparency and confidence-building
❑ Onsite Inspections	❑ Transparency and confidence-building	❑ Transparency and confidence-building	
❑ Transparency and confidence-building	❑ Notifications	❑ Destruction monitoring	
❑ Notifications	❑ Destruction monitoring	❑ Provisions for challenge inspections	
❑ Destruction Monitoring	❑ Provisions for challenge inspections		
❑ Provisions for challenge inspections			

| High | Ability to Verify | | Low |

Figure 6.1.   Arms Control Verification Comparisons

In considering the question of verification in arms control, a Congressional Research Service report identifies three objectives:[5]

- First, the regime should permit the countries to detect evidence that violations might have occurred.
- Second, the verification regime should deter violations to the treaty.
- Third, the verification regime should help build confidence in the viability of the arms control treaty.

If detecting, deterring, and building confidence in compliance is the desired outcome, then the accompanying verification mechanisms must do more than simply provide limits on weapons holdings and military activities. Transparency and openness mechanisms designed to improve the understanding of activities and to develop a dialogue are critical. This is really the basis for the development of tailored verification regimes, designed specifically for the types of systems and behaviors that are trying to be moderated.

That brings us back to the purpose of verification. While we have moved away from the concept of perfect verification in favor of effective verification, we must be mindful that a verification regime provides a set of agreed procedures that involves the collection of information and analysis of that info to allow judgments to be made as to whether a party is complying with its treaty obligations. Therefore, verification seeks to detect noncompliance, to deter would-be noncompliers, and to build confidence between parties while also enhancing transparency and openness. Perhaps as we think about verification, a critical distinction is to understand the important difference between monitoring that essentially provides a collection of information versus verification that is based on judgments of noncompliance.

When Nitze introduced the concept of effective verification, he did not address the costs of verification. Appreciation for the costs of verification include both financial costs (buying recon satellites, installing seismometers, developing new sensors, hiring personnel for on-site inspections, etc.) and the potential for compromise of our own secrets (including both confidential business information and government national security information unrelated to the subject matter of the treaty). Other important costs to consider are the potential implications for individual privacy and for false positives. Both have implications for the type of verification regime that will be acceptable.

He also did not address the question of enforcement; that is, once we've detected a violation, is there any legal mechanism (in the treaty or more generally) that can compel the violator to return to compliance? However, with his testimony on the INF treaty, Nitze moved the U.S. national security community away from an insistence upon "perfect" verification—the notion that we should not join the treaty if there was any realistic possibility of an undetected violation.

The benefits to be gained by an appropriate verification regime "sized" to provide a better understanding of the level of compliance, increase certainty in our judgments, and increase confidence in compliance clearly has important implications for any arms control agreement, which must be carefully balanced with the costs of such a verification regime.

The United Nations Disarmament Commission (UNDC) provided a list of sixteen "Principals of Verification" that gives yet another look at the issue (See Appendix E). The list is instructive for two reasons. First, it is quite thorough and specific in requirements for verification. Second, review of the principals suggests the degree to which the BWC has not followed the "approved format" for arms control treaties. For example, the language includes the caution that "verification arrangements should be addressed at the outset and at every stage of the negotiations." Obviously this guidance has not been followed to the letter for the BWC.

Still, the principles from the UNDC do serve as a good reminder of the relationship between arms control treaties and a viable manner in which to assess compliance.

HISTORICAL PERSPECTIVE: VERIFICATION OF BIOLOGICAL CAPABILITIES

The section above was intended to broadly consider arms control verification to allow the reader to gain a better understanding of verification in general. Little specific discussion was provided concerning the potential for verification of the BWC or the historical perspective of what we have learned about this issue. However, several important data points provide useful insights into the BWC verification protocol debate.

Understanding the history of national offensive BW programs has proven to be a challenge. In fact, even for programs that are clearly nascent, gaining a full accounting has proven difficult. The norms against the use of BW, the lineage of BW programs, some of the testing that may have been conducted, and the relationship between BW and special operations in many countries has combined to cause these programs to largely be shrouded in secrecy.

Soviet Noncompliance

The cases of Soviet BWC noncompliance discussed previously, Sverdlovsk and the "yellow rain" incident, demonstrate the difficulty of "proving" allegations of cheating on the BWC. In the case of Sverdlovsk, the inadvertent release of milled anthrax spores from a Soviet bioweapons factory occurred in 1979, was brought to the attention of the international community shortly thereafter, but was not admitted to have been caused by the Soviet offensive BW program until 1992 when President Boris Yeltsin provided an accounting of the incident. The "yellow rain" incident remains a mystery. Evidence suggests that a mycotoxin BW weapon was used in Southeast Asia, but no definitive information has been provided that establishes the cause-and-effect relationship.

More broadly, the BWC has not been effective in serving as a norm against the Soviet, and later Russian, pursuit of biological weapons. In fact, at the same time that the BWC was being negotiated, the Soviet Union was developing the largest, most sophisticated offensive BW program the world had ever witnessed. In 1973, the Soviet Central Committee created a new civilian agency, Biopreparat, within the Ministry of Health under the project code-name of Ferment for the purpose of developing an offensive BW capability. At the height of the program, Biopreparat had grown to over forty facilities and almost sixty thousand personnel. The Ministry of Defense also maintained three biological research institutes and a large, open-air testing facility.[6]

Taken together, these historical incidents—Sverdlovsk, "yellow rain," and the development of the Soviet offensive BW program—point toward inherent difficulties in developing an effective verification mechanism for the BWC. Much of what we know was actually provided by defectors who were able to provide important details of the Soviet program. Intelligence collection efforts were also important to piecing together the details; however, they were not definitive in establishing Soviet BWC noncompliance. Despite allegations and suspicions, short of an admission, proving that a BW incident has occurred has proven to be problematic.

In fact, the key role defectors have played in peeling back the details of the Soviet program has been a common theme for gaining an understanding of other national offensive BW programs.

Tripartite Agreement

The mounting evidence of Soviet noncompliance provided impetus for then U.S. President George Bush and British Prime Minister Margaret Thatcher to confront Soviet President Gorbachev. Gorbachev agreed to a series of joint visit facilities and a trilateral agreement designed to resolve the serious compliance issues that had been identified.

The first facility visits were held in January 1991. In this initial round, four Soviet facilities were visited, including the Institute of Immunology in Chekhov, the Institute of Microbiology in Obolensk, the Institute of Molecular Biology in Koltsova, and the Institute of Ultrapure Preparations in Leningrad. The locations were based on information from the first defector from the Soviet offensive BW program, Vladimir Pasechnik, who defected in 1989 and had worked at the Leningrad facility.[7] The inspections did little to alleviate the compliance concerns. They actually had the opposite effect of providing evidence of noncompliance and raising concerns about the size, activities, and scope of the Soviet and later Russian BW program. At Obolensk, the team was denied access to the aerosol test chamber and the plague research laboratory. At Koltsov, the team was confronted with the knowledge that smallpox research was being conducted. At Leningrad, explosive test chambers and milling machines provided evidence of a current offensive program. The visit to the four American sites in December 1991—the Soviets did not ask to visit any United Kingdom sites during this first round—were uneventful and showed a former U.S. program that was no longer operational. The visits were to Dugway Proving Ground, the United States Army Medical Research Institute for Infectious Diseases (USAMRIID) at

Fort Detrick, the National Center for Toxicological Research at Pine Bluff, and the Salk Institute, Government Services Division. While it was clear that that U.S. facilities, with the exception of Dugway that had an active weapons grid for aerosol testing of chemical weapons, were not operational, the Russian team expressed "concern" about the United States' mothballed capability. However, Ken Alibek, one of the Soviet defectors and the deputy at Biopreparat who was on the delegation in 1991, stated in his book *Biohazard* that the Russian delegation was able to see that the sites they visited were no longer used for offensive BW purposes.[8]

To formalize the inspection process and ensure that it would continue, the parties signed a trilateral agreement in September 1992. In a joint statement, the three governments confirmed their commitment "to the full compliance with the BWC and stated their agreement that biological weapons have no place in their Armed Forces."[9] The agreement set the stage for the continuation of the site visits as well as establishing working groups to consider visits to nonmilitary sites, cooperation in biological defense, conversion of facilities, exchange of information, periodic reports to legislatures and publics, and the exchange of scientists. The agreement further established the requirement for visited facilities to provide a standard set of information including briefings about the site, the site staffing, and information on hazards, safety, and medical requirements. In a sense, the tripartite agreement had created a mini-BW verification framework outside of the purview of the BWC.

Resulting from the agreement, the site visits continued. The next four included the All-Union Scientific Research Institute of Veterinary Virology at Pokrov, the chemical plant at Berdsk, the chemical plant at Omutninsk, and the All-Union Scientific Research Institute of Microbiology at Obolensk for a second visit. Three of the four facilities provided evidence of offensive BW work since 1975 when the BWC entered into force. Reciprocal visits were held in February 1994 in the United States at Pfizer U.S. Pharmaceuticals in Vigo, Indiana, and Groton, Connecticut, and Plum Island Animal Disease Center in New York (formerly at a Department of Agriculture facility and now a Department of Homeland Security facility that supports the USDA), and the Evans Medical Limited in Liverpool in the United Kingdom. The visits to the U.S. and U.K. facilities provided no evidence of offensive BW, but it did cause serious concerns among the commercial companies being inspected. Interestingly, Pfizer was so concerned about protection of its commercial interests that Vice President Al Gore had to intercede before allowing the visit to occur.[10]

After the 1994 visits, additional negotiations between the three parties continued, but over the next two years, they were unable to reach agreement on the continuation of the activity. These meetings included "rules of the road" discussions intended to develop a framework for future site visits including defining the selection of sites, operational procedures for the visits, time limits, and mutually agreed objectives. They also were intended to serve as forums for exchanging information on previous BW activities. The United States and United Kingdom provided a comprehensive report; however, the Russians did not, only providing far less information than had been hoped for and calling into question their openness and transparency. After 1996, all activity related to the trilateral agreement ceased and the program became inactive.

Interestingly, while access to the facilities was actually quite good with the exception of a couple of areas within some of the Russian facilities, it was not possible to establish unequivocally that the Russians were not in compliance with the BWC. Certainly, evidence existed that proved that work on biological weapons had been conducted since 1975, but determining whether the work was defense and therefore in compliance with the BWC or offensive and in violation of the BWC was not possible. Certainly accounts suggest that the U.S. and U.K. inspectors believed the work was part of a continuing offensive BW program; however, that assessment could not be definitively concluded.

In a very real sense, the Tripartite served as a harbinger of the difficulties associated with developing a BWC verification protocol. Lessons learned included:

(1) an accused party may react strongly to an allegation of noncompliance, demanding strict reciprocity; (2) the importance of intelligence information for the planning and conduct of on-site visits; (3) that there is a clear value in an unambiguous account of past BW activity being provided; (4) that clear technical objectives are required for an effective inspection regime; (5) that true short-notice inspections are difficult to achieve; (6) that the redeployment of BW program personnel to legitimate civilian activity needs to be accomplished; (7) technical assessments of observations made during site visits may have been influenced by political interpretations based on other intelligence; and (8) team composition and selection are very important.[11]

By way of a postscript to this section, a 2011 Department of State arms control compliance assessment writes, "It also remains unclear if Russia has fulfilled its obligations under Article II of the BWC to destroy or divert to peaceful purposes the items specified in Article I."[12]

Overall, the reciprocal visits conducted between 1991 and 1994 had two rather chilling, long-term effects. First was related to the inability to establish with certainty the nature of the activity being conducted at the sites. To the United States, this implied that verification of a BWC through a formal verification protocol was problematic. Second were the concerns expressed by the commercial sector about the disruptive nature of the visits and the concern for industrial espionage.

UNSCOM Inspections in Iraq

In August 1990, Iraq invaded the Kingdom of Kuwait in a complex air, land, and sea operation. Over the course of the next nine months, a coalition first defended the territorial integrity of Saudi Arabia and then attacked to expel the forces of Saddam Hussein out of Kuwait. The terms of the cease-fire agreement included provisions for establishing a mechanism to ensure the elimination of Iraq's weapons of mass destruction. To this end, on April 1991, the United Nations Security Council established UNSCR 687.

Iraq had been a 1972 signatory to the BWC but did not ratify the treaty until 1991 after its defeat and the establishment of UNSCR 687. Given this status, Iraq was not in technical violation of its responsibilities under the Convention during the period from 1972 and 1991 when it was developing a BW capability.

One of the provisions was the establishment of the Special Commission or UNSCOM to oversee this WMD disarmament, including Iraq's biological program. Over a five-year period, the UNSCOM team came to know much about the Iraqi program, seeing the original 1992 BW declaration go from seventeen pages to 639 pages in the 1997 declaration.[13]

The initial meetings and engagements were filled with "denials and falsified documents."[14] Saddam had hoped that the UNSCOM mission would be short-lived and that he could wait out the inspectors. This proved to be a false hope, as the team set about its task of mapping and then establishing monitoring capabilities for Iraq's laboratories, biological institutions, and medical facilities.

Some scant evidence of Saddam's biological programs was known in the period just prior to the invasion of Kuwait, but far more was known about Iraq's chemical program that he had used in the Iran-Iraq regional war in the 1980s. Periodically, reports surfaced about a biological program, and several agents including anthrax, plague, and botulism were identified as potential pathogens. However, direct evidence of such activities had not been definitively established. Still, concern about biological weapons caused U.S. and coalition planners to bomb suspected BW sites and provide vaccinations against several potential biological agents.

In the aftermath of the Gulf War, UNSCOM inspectors conducted a series of inspections to find the smoking gun, the evidence that the Iraqis had engaged in a secret bioweapons program. For almost four years following the war, UNSCOM engaged in a cat-and-mouse coordinated effort by the government of Iraq to conceal the facts behind the work that they were doing with biological material.[15] The Iraqis initially claimed that all efforts were directed toward civilian applications in areas such as biopesticides and vaccine development. Along the way, many anomalies were uncovered, such as the large amount of growth media that had been purchased for unknown reasons, the lack of containment around facilities that likely were used for offense BW production, and the many odd coincidences such as having strains of brucellosis and tularemia that had been developed by other nations for experimentation for use as BW weapons.[16]

Cameras had been installed in some of the facilities, multiple visits to sites had been conducted, and even residue tests had been conducted that were inconclusive. A former commander of USAMRIID, David Huxsoll, stated, "There was never a question in my mind as to what they were up to," after visiting two suspected BW sites. Still, there was no proof of violations of the BWC.

It is telling, however, that after almost two years of inspections, discussions with Iraqi scientists, and investigation, the UNSCOM team composed of experts in the field disagreed about whether Iraq had a bioweapons program.[17] By 1995, the assessment still was not definitive. One account from the leader of the UNSCOM team, Rolf Ekeus, stated, "They have a capability which we have not as yet put our hands on fully; a capability to mass-produce microbiologic items—viruses and bacteria—for warfare purposes."[18] Still, the statement stops short of declaring that an offensive BW program had been discovered.

In June 1995, the UNSCOM team took their findings to the UNSC. The statement had to be carefully worded as there were plenty of suspicions that an offensive program had existed, but the evidence was circumstantial and the Iraqis had not admitted to such a program. In a meeting just prior to the release of the report, the head of the team met with Tariq Aziz, the Iraqi foreign minister, to preview the findings and say that the cooperation had been less than full and anomalies had been discovered that raised serious questions about Iraq's offensive biological program. Following this meeting, Ekeus went back to Iraq where he would receive the first official admission that the Iraqis had developed an offensive BW capability. The admission was only partial as later additional admissions would be forthcoming.

Two months later, Lieutenant General Hussein Kamal, Saddam's son-in-law, a former Minister of Defense, and the current head of the Special Security Organization, defected. He was certainly knowledgeable about the Iraqi program and gave some details that were useful as confirmatory statements. However, the most important role Kamal played in providing information on the Iraqi program was based on the additional admissions that were made by the Iraqis in the process of blaming him for directing the efforts of the secret program. In the immediate aftermath of his defection, Iraq admitted to having experimented with other pathogens in addition to anthrax, plague, and botulinum neurotoxin to include *Clostridium perfringens* (or gas gangrene) and aflatoxin. Additionally, whereas previously Iraq had only admitted to having bulk storage of the pathogens, now they admitted to having rockets, warheads, and bomblets that had been filled with weaponized pathogens. As a footnote, some attribute the defection of Kamal as being instrumental to Iraq's final unequivocal admission. Perhaps, but it was clear that after four years of manipulation and false statements, the Iraqis had run out of plausible deniability.

The UNSCOM experience in Iraq holds great importance to the question of verification. Even with extraordinary access—including "unrestricted freedom of movement without advance notice within Iraq of the personnel of the Special Commission and its equipment and means of transport"—certainly more than would be granted under a BWC verification protocol, the task of unraveling the Iraqi BW program took years of effort, hundreds of thousands of man-hours, and access to a high-level defector.

Assessing Noncompliance

The tripartite agreement visits and experiences in Iraq demonstrate that determining compliance with the BWC is a difficult task even if given the opportunity for site visits and discussions with scientific personnel. Assessing compliance from afar is even more problematic.

On an annual basis the U.S. Department of State provides assessments of the compliance of a range of international arms control agreements. For each of the agreements, the document provides a short introduction and then country-specific assessments highlighting the degree of compliance assessed and concerns noted.

For one of the annual reports, the difficulty of reaching conclusions about BWC compliance is expressed as

> The emphasis in Article I upon whether or not the purposes for which materials or items are possessed are "peaceful" or "hostile" makes clear that not only the existence, but also the intent of any country's biological program, must be part of an compliance determination. Making a judgment about intent is challenging given the dual-use nature of most biotechnology equipment, facilities, and activities. As with other agreements—particularly those involving dual-use technologies that may be used in a variety of legitimate or illegitimate ways—intent is a critical element, and it may have to be inferred from the circumstances, in light of all available information, if direct evidence is not available. [19]

The county-specific assessments do just as articulated in the paragraph above. That is, they make assessments based on inferences and the best available information. While not intended to be a criticism, this methodology is far from scientific and certainly not able to be replicated. In short, it is based on judgments rather than facts.

Confidence Building, BWC Style

The confidence-building measures (CBMs) that were started in the Second Review Conference and refined in the Third Review Conference serve as an interesting case study in this verification debate.

Any verification protocol that would be adopted would begin with a data submission. The eight data fields required under the current CBMs provide such a format for this sort of submission. However, many have questioned the utility of the current CBM format as a means of understanding BWC compliance. This has initiated an effort to revise the CBM format. Additionally, experience indicates that only seventy or so nations out of the 168 member states parties of the Convention submit the voluntary national CBM submissions. The annual submissions are provided to the ISU for inclusion on the internal website. Several nations, approximately twelve or so each year, have decided to have open submissions that are posted for all to view. In 2010, the United States posted its submission for 2009 on the public website.

In trying to assess why there is a low submission rate, several scenarios are possible. Perhaps nations do not think that they have any submission to make, as they have not ever had a BW program. Perhaps they see the submission as overly intrusive. Perhaps they find the submission overly complex. Clearly, cursory examination indicates that all nations should be making submissions, if only to report national laws and regulations dealing with biodefense issues. For whatever reason, the lack of universal participation in the BWC is indicative of the malaise that surrounds the Convention and requires immediate international attention. A contributing factor is certainly the lack of resourcing for the ISU; with only a three-person organization, conducting outreach to encourage broader participation will remain problematic.

This is not to imply that the current CBM format is adequate for verification, but rather that the support for even an annual data submission does not appear to be present, much less for a verification regime that would entail considerably more time and attention.

THE LIMITATIONS OF VERIFICATION

The above discussions of historical verification examples from biological programs provide an indication of the complexity surrounding the development of a verification regime. By way of a conclusion to this chapter, several points are worthy of review.

First, verification is by its very nature judgment based. There are no absolutes. Even with significant resources and access, making judgments about activities that are ongoing are complex and shrouded in uncertainty.

Second, even teams of highly trained and prepared individuals are likely to come to different conclusions when conducting inspections of biological facilities. The dual-use nature of biotechnology makes reaching definitive conclusions very difficult.

Third, no one single technique will suffice for verification. It is the combination of discussions with scientists, on-site visits, intelligence, and expert analysis that allowed the UNSCOM team to finally establish that Iraq had an offensive BW program.

Fourth, the value of an admission by the state in question can be imperative to establishing unequivocally that a state has had an offensive program.

Finally, there is the question of verification versus compliance. As the definition at the beginning of the chapter reminds us, "Verification is the process that one country uses to assess whether another country is complying with an arms control agreement. To verify compliance, a country must determine whether the forces or activities of another country are within the bounds established by the limits and obligations in the agreement." Compliance on the other hand is establishing whether a nation is fulfilling its obligations under the treaty. A state can be in compliance, yet its compliance is not verified. However, the reverse should not be possible; that is, a state should not be able to be verified as meeting its obligations if the state is not in compliance.

Chapter Seven

The Future of the BWC

As we think about the future of BWC, a recapitulation of the effectiveness of the Convention is in order. Along with this assessment, we need to examine what we desire from the BWC in the long term. What are the threats that drive the future of the Convention? How do we balance the surge of biotechnology with the need for appropriate security? What are we trying to accomplish with the BWC? Control technology? Ensure a level international playing field? Are the outcomes that we seek even feasible in this Age of Biotechnology?

The desired outcomes for many of these questions are likely to vary greatly among countries, from developed to developing, from the Western Group to the Non-Aligned Movement, and even among close allies. Being able to triangulate these different positions and find a solution space that accommodates the range of concerns and desires will likely be very tricky.

In looking at the question of the effectiveness of the BWC, one must consider the relatively few man-made biological incidents that have occurred post-1975. Since the BWC entry into force, there have been less than ten incidents of use, of which none have been demonstrated to have been caused directly by state biological programs.[1] All biological incidents have been attributed to lone actors or terrorist groups. No linkages to state programs have been established with the exception of the Amerithrax case in the United States, where the alleged perpetrator directly obtained the seed stock from the U.S. defensive biological program. In some other bioterrorist incidents, the perpetrators have obtained pathogen samples from legitimate commercial sources.

This is not to imply that the states have not been involved in questionable programs that have exceeded or at least approached what the BWC would likely consider reasonable for a defensive program. Still, there does appear to be a threshold regarding the use of BW that states have been unwilling to cross.

Given this history, we must then ask the pointed question of whether the BWC, with all its warts—lack of a verification protocol, lack of universal adherence, reliance on a voluntary data submission to name a few—is good enough, or on the contrary, should more be done to address the shortfalls that are inherent in the BWC? We must ask the hard question of whether pushing too hard on this fragile Convention could do more harm than good. Would pushing too hard on verification, for example, result in a fracturing of the BWC?

My position on the BWC will come as no surprise to the reader given the title of the book. However, it is worth restating. The BWC is the most important arms control agreement of the twenty-first century, with direct linkages across the breadth of human activity. We must continue to ensure the viability and relevance of the Convention at all costs. Further, we must find linkages between other existing international forums and the BWC. While organizations such as the WHO, FAO, and OIE have relevance across human, agriculture, and plant issues, they deal primarily with the range of nonsecurity issues. Only the BWC is exclusively focused on security.

Let's now look at the elements of the future BWC that will serve us well into the twenty-first century.

UNDERSTANDING THE THREAT

Historically, arms control agreements have been established to limit holdings of weapons systems, to control behavior, and to increase transparency and stability. As such, they are directly related to perceived threats, risks, and concerns of the parties to the treaties.

When the BWC was negotiated, the major concerns related to the gaps resulting from the 1925 Protocol for the Prohibition of the Use in War of Asphyxiating, Poisonous or Other Gases, and of Bacteriological Methods of Warfare. The use of biological and chemical weapons by states was seen as an abhorrent form of warfare that should be eliminated. The primary threat was considered to be states developing a biological weapons capability to use against other states. The notion that a nonstate actor would use such techniques was not considered to be a behavior to be limited by this Convention. It is not that there was no precedent for this type of activity; for example, there was the use of *Clostridium botulinum* in 1910 by Pancho Villa, a nationalist separatist against Mexican federalist troops.

The timing of the BWC negotiation was also highly relevant as it was at the height of the Cold War, when the globe was divided between East and West. The balance of power was essentially bipolar, with the United States and the Soviet Union and their direct allies pitted against each other. The standoff was punctuated by the nuclear philosophies that saw the two super-powers essentially extend their nuclear umbrellas over their alliances, NATO and the Warsaw Pact. Even nonaligned nations became fodder for the stand-off, resulting in a string of proxy wars in Africa, Latin America, and Asia.

So in 1975, the state threat of the use of BW drove the Convention in the direction of a typical arms control agreement between states parties. Today, our experiences indicate that the range of threats has evolved. We continue to have concerns about state BW programs, but we have also recognized a spectrum of biological threats that begins on one end of the continuum with state threats and naturally occurring disease on the other. In between are activities such as the misuse of technology, accidents, and negligence.

This might seem like a trivial issue, but the lack of agreement on the threat has resulted in differences in national approaches to implementation of the Convention. Those nations that see the threat as predominantly coming from state actors will be more likely to view issues such as verification through the prism of having to constrain state BW efforts and therefore rely on an international verification regime. On the other hand, if a state sees the issue as predominantly a bioterror issue or requiring consideration of the full spectrum of threats from naturally occurring disease to state BW threats, then the perspective on issues such as verification will be more focused on issues such as national implementation.

Looking at the question of CBRN issues more broadly, experts generally agree that these weapons issues are low-probability, high-consequence events. Fissile material suitable for use in a nuclear weapon is difficult to acquire. Acquiring a nuclear weapon would likely be even more difficult. Radiological material is easier to acquire, but it is unlikely to be seen as a high-consequence event. Generally, chemicals such as chlorine and ammonia that would be suitable for use in an attack would be easier to obtain than acquiring fissile material, yet experience indicates that very large amounts of chemicals would be required to have a high-consequence outcome. Develop-ing more potent chemicals such as sarin nerve agent requires more skill, knowledge, and precursor chemicals that are highly controlled; therefore, one cannot reasonably consider them as high probability.

Natural Disease Outbreak	Unintended Consequences	Accidents	Negligence	Vandalism, Sabotage	Deliberate Use of BW

Figure 7.1. Spectrum of Biological Threats

In contrast, the potential for biological threats varies from low probability, low consequence to high probability, high consequence. The spectrum of biological threats encompasses a broad range of biological incidents from naturally occurring pandemics such as influenza to the state use of BW.

Let's look at some biological threat examples. A low-probability, low-consequence biological event would include the 1996 contamination of pastries with *Shigella dysenteriae*, the patient with multi-drug-resistant tuberculosis traveling on a commercial airliner, limited outbreaks of influenza H7 in poultry, or laboratory-acquired infections from BSAT organisms. The Amerithrax attacks of 2001, which resulted in only twenty-two cases and five deaths but had high economic and social consequences can be considered to be a low-probability, medium-consequence event. Meanwhile, the 2009 H1N1 influenza could arguably be considered a high-probability, medium-consequence event. Had this pandemic been caused by a highly pathogenic strain of influenza such as the H5N1 avian influenza (with a 60 percent mortality) with the transmissibility of the standard seasonal influenza, it would undoubtedly be considered to be a high-consequence influenza event.

Of course, the definition of low consequence would be very subjective. Some might argue that the Amerithrax attacks that resulted in an increase in biodefense funding from $400 million to over $7 billion per year in nonmilitary biodefense should be seen as high consequence. Still, the argument that biological threats run the gambit from low probability, low consequence to high probability, high consequence is a reasonable assertion.

In thinking about the importance of understanding the threat to the BWC, where one starts is critical. If one approaches the issue as a state threat, the natural progression is an international verification regime. However, if the threat is perceived to be primarily from a bioterrorist, then one gravitates toward a system of national implementation augmented with the appropriate international hooks to handle a state BW issue of noncompliance or attack.

Examining the changes that are underway in biotechnology, one must conclude two important points. First, biotechnology is experiencing extraordinary growth that has the potential to alter life on this planet. We are seeing a convergence of several technologies including chemistry, informatics, and engineering with biology that will change the world dramatically from the food that we consume to the energy sources we use to our very quality of life and longevity. Virtually every part of the human existence will be affected by biotechnology. Second, the thresholds for entry into the field of biotechnology have been greatly lowered over the past forty years since the negotiation of the BWC and these trends will undoubtedly continue well into the future. This will make biotechnology more accessible to a broader cross-section of the global population, which will increase the likelihood of biological accidents and misuse.

THE EFFECTIVENESS OF THE BWC

In looking at the future of the BWC, a useful starting point is the question of the effectiveness of the current Convention and its provisions in deterring states from developing, stockpiling, transferring, and the use of biological weapons.

By these measures, one would have to say that the BWC gets mixed reviews. After the entry into force of the BWC, the number of states known or thought to have offensive biological warfare programs has actually increased. In fact, the massive Soviet BW program actually accelerated following the negotiation of the 1972 accord. This is ironic given that the Soviet Union was one of the three Depository Governments along with the United States and the United Kingdom.

On the issue of the development of offensive biological capabilities, one would be tempted to conclude that the treaty has failed. However, if one then considers the details of the likely programs, a different picture emerges. For example, of the states that are thought to have had or potentially have a BW capability, few were thought to have actually weaponized the material, which is the linking of the pathogen with a delivery system such that it could be used as a weapon. Most of these efforts were assessed to be developmental more akin to research and development than weapons development. This is not meant to excuse those nations that might have cheated and therefore, violated their obligations under the BWC, but rather to understand the type of behavior that these nations engaged in. Even fewer were assessed to have stockpiled quantities of biological weapons.

Coming back to the CBRN analogy, it appears that many nations see the norms against the use of biological warfare—the deliberate spread of disease—as even being less acceptable to global society than the other elements of CBRN. Therefore, when one takes this question of BWC relevance and development of BW capabilities to the next level, no states have overtly used biological weapons as a deterrent. This is in contrast to nuclear and chemical weapons, in which the employment doctrine was stated and even factored into the deterrence posture and warfighting capabilities of several nations, such as the United States, Soviet Union, and France, to name a few.

Of course, since the Convention entered into force, some allegations of cheating have been made, but they are largely unproven except in the cases of the Sverdlovsk incident and several notable assassination attempts, including Georgi Markov, a Bulgarian writer who was killed on a London street when a pellet containing ricin was fired into his leg via an umbrella by the Bulgarian secret police.

Given the professed killing and maiming power of biological weapons, a puzzling question then is why nations have not chosen to use BW for deterrence and have largely kept thoughts about the use of biological weapons as state secrets. Perhaps it is the fear of disease that one author notes concerning the use of disease as a form of warfare. Perhaps it is the unequivocal norm that has been established concerning the use of biological weapons. It is also very curious that given the relative ease with which biological weapons could be developed that we have not seen more interest in BW as a form of asymmetric warfare by nonstate actors. This is not to say that no nonstate actors, terrorist groups, or lone wolves have expressed an interest in developing and even using such capabilities, but the numbers are surprisingly low and the attempts to use BW weapons are so infrequent as to suggest little overall interest in this form of "warfare."

Two questions continuously arise when considering the effectiveness of the BWC. First, can a treaty be effective if it has no verification regime or definitions? Second, how can one judge whether a violation has occurred? Both are worthy of additional discussion.

Concerning the question of verification and definitions, historical evidence, as discussed previously, suggests that the BWC has been an effective norm against the use of biologic weapons. On the other hand, the large number of nations, including several of our close allies, believe that a verification regime and the definitions that would accompany such a mechanism suggest that over time it will need to be considered. Of course, any changes must be considered within the overall effectiveness, or perhaps the perceived effectiveness, of the Convention that could be diminished if not handled appropriately. We have already seen from the results of the 2011 review conference that significant dissatisfaction has been expressed over the lack of such verification or compliance mechanisms and the unwillingness of the United States to engage in any sort of dialogue that would imply movement toward a verification protocol.

The final days of the 2011 review conference also provide insights into the degree to which the verification issue has and will continue to be a source of irritation for many. Whether the issue is real or perceived as a way to irritate the United States is debatable, but the effect on the BWC is real. Certainly, nations such as Russia have been having the best of both worlds, with being able to blame the United States for the lack of a verification protocol while understanding that such a protocol would involve opening up several of their sensitive sites.

One assessment of the review conference indicates the precarious balance of the BWC for failing to make progress on the important biological issues of the day.

The December 2011 review conference of the Biological Weapons Convention (BWC) demonstrated the danger of the bioweapons ban drifting into irrelevance. Standstill was the motto of the meeting. Only incremental improvements on some procedural issues were achieved.[2]

Of course, one of the most significant issues in this progression toward irrelevance is the question of verification, which consumes a great deal of energy but achieves little.

On the second issue concerning assessing whether a violation has occurred, this is both directly related to the verification issue and also one of the major reasons why such a regime is of such concern to the United States. The idea of lengthy investigations that would likely prove inconclusive, the resulting loss of security and sovereignty, and the potential economic espionage have combined to make the very idea of a verification regime unpalatable to the United States. Still, we must acknowledge that processes do exist for assessing whether violations of the Convention have occurred, but the assessments will be largely based on subjective assessments rather than objective measures. Even when a smoking gun was found during UNSCOM inspections in Iraq and with the extraordinary access that was granted that would surely not be available through a normal verification protocol, definitely establishing that violations had occurred was very difficult until the Iraqis confessed to their transgressions.

A DIFFERENT KIND OF TREATY

Arms control, by the very nature of the topic, is a security-based subject. It involves strong elements of national interest, and the avoidance of activities that are of mutual concern. Additionally, it results in minimizing the costs and risks of the arms competition, and limits the scope and violence of war in the event that it occurs.

The dialogue surrounding the development of the Convention certainly provides evidence that the framers understood that it would not be possible to transplant nuclear arms control frameworks into the biological warfare debate. In this way, it is clear that the nations negotiating the language for what would become the BWC recognized that the Convention was unique; it was a very different type of international agreement. Unlike other arms control treaties that included strict definitions and requirements for verification, the BWC was a short statement of political purpose that called for the ban on all offensive biological efforts.

Perhaps the genius of the BWC was the understanding that the Convention had direct applicability across one of the fastest-changing fields in human history and that attempting to develop a standard arms control framework would not be viable.

These differences have continued to grow as the field of biotechnology has continued to evolve. Whereas typical arms control agreements such as the Nuclear Non-Proliferation Treaty (NPT) or even the New START Treaty focus on weapons and capabilities that are the exclusive purview of a state, the BWC has wide application across multiple constituencies including public health, medicine, security and defense, agriculture, environmental issues, food security, and biodiversity.

Even the CWC, which limits chemical capabilities, pales by comparison to the BWC in its daily intersection with peoples and societies. This is largely because the chemicals that have most relevance for use as weapons are highly controlled, including the precursors required for the manufacture of these weapons. Additionally, key processing is required to make chemical weapons. This is not to imply that a chemical attack using even everyday pesticides could not be effective. An attack of this sort could sicken and kill, but it would be confined to the area of release and would likely have a tactical outcome, depending on the amount of chemicals released.

The same is not true for biological weapons where pathogens with virtually no processing could be used in a BW attack. Perhaps they would not be as effective as if the agent had been specially processed to make it more stable in the environment, resistant to drugs, or prepared for an aerosol delivery. However, even relatively rudimentary preparations have the potential to cause high mortality and morbidity, given the proper pathogen. Imagine the potential panic and response should smallpox be reintroduced into the public domain or if a similar incident to the Rajneeshee attack on the salad bar used Ebola virus rather than the Salmonella bacterium that causes salmonella.

These are areas where the governments that are parties to the BWC do not have exclusive capabilities. In fact, for most of the member state parties to the Convention, the government lags well behind advances that are occurring in biotechnology. This is certainly true in the United States and other nations with strong biotechnology industries. Consider the development of the first synthetic virus by Dr. Eckard Wimmer in 2002, as he and his team were the first to synthesize in a test tube a virus (poliovirus); the work done by Dr. Craig Venter on rapid genomic sequencing and producing bacteria that have been engineered to perform specific reactions, for example, producing fuels and making medicines; the emergence of bioinformatics that is fundamentally altering our knowledge of genomics and rational drug design; and even such technologies as the aerosol delivery of drugs and vaccines such as insulin and influenza. All of these advances have what we have taken to calling dual-use applications. That is, they are capabilities that have been developed for legitimate purposes that have direct application for use in developing biological weapons. No other arms control treaty even comes close to the impact of the BWC on the lives of citizens and societies.

The sheer magnitude of the biotechnology field in the United States also provides evidence of our significant biodefense program and linkages with other related fields. A 2007 Government Accountability Office (GAO) report listed fifteen biosafety level (BSL)-4 and 1,356 BSL-3 laboratories that were registered with the USDA and CDC. This only reflects the total number of these high containment facilities in the United States.[3] When one considers that a large percentage of the pathogens for potential use as bioweapons could be handled safely in a BSL-3 facility, the magnitude of the issue comes into clearer focus. Only a handful of the most deadly pathogens for which there are no known countermeasures or vaccines are classified as BSL-4. This is not to imply that all BSL-3 and -4 facilities are working on select agents (SA), those agents that have the most direct applicability for use as bioweapons. On the contrary, the biosafety levels have been established not to secure the pathogens but rather to protect the scientists and technicians from exposure to dangerous pathogens during the course of their work. Biosafety levels also relate to the protections taken to prevent the inadvertent release of a pathogen from the facility where the work is ongoing. It is worthy of note that no comprehensive regime or oversight agency exists for BSL-3; only if select agents are to be used does the facility require periodic inspections and a higher level of scrutiny.

This is where the dual-use picture comes into focus. We want scientists to work on pathogens in the proper containment levels, but we also realize that doing so provides a potential proliferation window should the pathogens escape the facility or be deliberately removed. If there is a highly virulent pathogen and little is known about it, prudence mandates that appropriate measures be taken. The 2011 outbreak of the Schmallenberg virus in Germany—which causes congenital malformations and stillbirths in cattle, sheep, and goats—had previously not been identified, and therefore high-containment facilities were used for the initial evaluations. After more is known about the pathogen, the level of containment (as measured by biosafety levels) can be adjusted if it is found not to require a high-containment facility. Key factors in determining the level of containment are the availability of medical countermeasures and the potential for high mortality and morbidity.

For this reason, we must treat the BWC as a special class of arms control treaty and be extremely cautious about constructing requirements and regimes that would have an adverse effect on legitimate scientific discovery and advances in biotechnology that will undoubtedly lead to significant enhancements in the quality of life of the global community.

UNIVERSAL ADHERENCE

The lack of universal adherence to the Convention has been and will continue to be a nagging question. Despite these concerns, only modest progress has been made in rectifying the situation. Without a greater international commitment, the BWC will remain irrelevant to a large percentage of the global community.

The number of member states has grown to 168 member nations. This reflects a steady increase. However, when compared to the NPT, which governs nuclear proliferation and has 189 members, and the CWC that has 188 members, the BWC remains undersubscribed with its 168 members. This is a complaint that is mentioned frequently.

Along with the issue of the number of member state parties, we should also be concerned with attendance at the annual meeting of state parties and the review conferences. In this regard, what is more damning is the lack of participation at the 2011 BWC Review Conference. A total of 103 nations of the 168 member states participated in the 2011 review conference. Several other states and more than fifty nongovernmental organizations and UN staff elements participated in the RevCon under observer status.[4] To rectify this situation, the BWC community must find a path to involving more member nations in the regular meetings as a matter of the highest priority.

However, attendance is not the only measure that indicates undersubscription to the Convention. The poor showing on the "voluntary" annual confidence-building measures clearly demonstrates certain lethargy with respect to the BWC. One would hope that over time the importance of the BWC would lead to greater participation; however, this has not been the case. During the intersessional period from 2007 to 2011 the number of national submissions ranged from sixty-three to seventy-two nations. In 2012, the total number of nations has actually declined to fifty-nine nations, or slightly less than 36 percent of the 168 nations that are now member state parties. These submissions are posted to the internal ISU webpage that is password protected. Some twelve nations or so have annually provided their submissions to the open website where it has been made available to the public. This now includes the United States, which began posting the national submission on the open website in 2010.

THE BWC IN THE AGE OF BIOTECHNOLOGY

In looking to the future of the BWC, one clearly senses that the Convention is at a strategic crossroads. The divide between developed and developing member nation states is widening. Instead of being a forum for cooperation, member states are using the BWC as a forum for advancing alternative, non-BWC-related agendas. More states are becoming members, yet the adherence to the CBMs and attendance at meetings has declined. At the same time, the potential impact of the treaty has never been greater, with rapid advances in biotechnology. These advances also suggest that the need for the BWC has never been greater. The debate about a verification protocol does not appear to be waning; in fact, if anything, the most recent review conference indicates a renewed desire for the protocol after the hiatus of 2006 following the failed 2001 to 2002 RevCon, where the United States walked out of the conference. So what should the BWC evolve into in this Age of Biotechnology?

In answering this question, several key areas remain priorities for the BWC of the twenty-first century. They include verification and compliance, the organization and structure of the BWC, assessing the impact of biotechnology on the Convention, sharing the benefits of technology, operational relevance of the Convention, and managing change of the BWC.

Changing the Dialogue

Any verification regime for the BWC must include by design both national and international components.

In considering the verification debate, one truism is that we must find a way to change the dialogue. Earlier, we introduced a definition of verification as "the process that one country uses to assess whether another country is complying with an arms control agreement. To verify compliance, a country must determine whether the forces or activities of another country are within the bounds established by the limits and obligations in the agreement."[5] It is interesting that the definition above contains both the terms *verification* and *compliance*. In this definition, the act of verification becomes the technical inspections and assessments, while determining compliance is about the judgment of a state's adherence to the agreement.

Paul Nitze, former Deputy Secretary of Defense, introduced the concept of "effective verification" as a means to determine whether "militarily significant" breaches had occurred. This moved the verification debate away from perfect verification to one of good enough. If one were to apply these same standards to the BWC, what would "effective verification" entail? Would a verification regime need to go to each of the fifteen BSL-4 laboratories in the United States? What about the 1,356 BSL-3 laboratories that operate around the United States and could easily be used for work with pathogens capable for use as biological weapons? What would it take for others to assess that the United States was in compliance with the Convention? Would there ever be enough evidence that some "cheating" or technical violations of the Convention were not being made? Perhaps the more relevant question is how many intrusive inspections and reports would be enough to declare that a nation was in compliance with the BWC?

In changing the nature of the debate, a more balanced threat and risk-based approach should be undertaken. The current verification protocol discussions are focused on gaining international assessments of national compliance. National implementation is treated as a separate and distinct issue. If one were to have a more holistic view of the threat as the full range of threats represented by the "Spectrum of Biological Threats," which goes from naturally occurring pandemic disease to state BW programs and includes accidents, negligence, and terrorist BW programs, the requirements for verification change. Instead of a narrow focus on assessing whether a state has an offensive BW program, the onus is on a state demonstrating compliance with all of the articles of the Convention against the full range of threats and risks.

In such a formulation, a nation that fails to have national laws, policies, and regulations governing the development, production, stockpiling, acquiring, or retention of biological weapons would be judged to be not in compliance with the obligations of the Convention. This is not meant to imply that a cookie-cutter approach would need to be implemented for developing national laws, regulations, and policies, but rather that some controls would need to be demonstrated in order to be judged to be in full compliance with the articles of the Convention.

In the United States, the documentation associated with control of biological capabilities and materials is extensive. Since the Amerithrax attacks in 2001 alone, a number of National Security and Homeland Security Presidential Decisions (NSPD and HSPD, later called Presidential Policy Directives [PPD] in the Obama Administration) policy documents, executive orders (EO), and laws have been established. Examples of some of the major documents are:

- HSPD-4 National Strategy to Combat Weapons of Mass Destruction (2002)
- HSPD-5 Management of Domestic Incidents (National Response Plan) (2003)
- HSPD-8 National Preparedness (2003)
- HSPD-9 Defense of Agriculture and Food (2004)
- HSPD-10 Biodefense for the 21st Century (2004)
- HSPD-10, A1 National Policy on Classified Life Sciences Research (2007)
- HSPD-18 Medical Countermeasures against Weapons of Mass Destruction (2007)
- HSPD-21 Public Health and Medical Preparedness (2007)
- EO 13527 Timely Provision of Medical Countermeasures (2009)
- EO 13486 Strengthening Laboratory Security in the United States (2009)
- EO 13546 Optimizing Security of Biological Select Agent and Toxins in the U.S. (2009)
- PPD-2 Implementation of National Strategy for Countering Biological Threats (2009)
- Proposed EO United States Government Policy for Institutional Oversight of Life Sciences Dual Use Research of Concern (2013)

The documents provide requirements, guidance, and policies across a broad range of activities that are directly related to the articles of the Convention. For example, HSPD-10, Biodefense for the 21st Century, published in 2004 during the George W. Bush Administration, contains four major areas of activity, including threat awareness, prevention and protection, surveillance and detection, and response and recovery. This document remains relevant as of this publication, serving as the most coherent strategy for the national biodefense mission.

A more recent document, PPD-2, serves as the implementation for President Obama's *National Strategy for Countering Biological Threats*, which was actually first publically released at the 2009 BWC meeting of states parties and provides seven objectives: (1) promote global health security; (2) reinforce norms of safe and responsible conduct; (3) obtain timely and accurate insight on current and emerging risks; (4) take reasonable steps to reduce the potential for exploitation; (5) expand our capability to prevent, attribute, and apprehend; (6) communicate effectively with all stakeholders; (7) transform the international dialogue on biological threats.[6]

The *National Strategy for Countering Biological Threats* also provides an important view of the U.S. position on biological issues, including the BWC, in stating,

This Strategy reflects the fact that the challenges presented by biological threats cannot be addressed by the Federal Government alone, and that planning and participation must include the full range of domestic and international partners. It is guided by the following assumptions:

- Advances in the life sciences solely used in a peaceful and beneficial manner should be globally available;
- A biological incident that results in mass casualties anywhere in the world increases the risk to all nations from biological threats;
- Biological attacks against animals or crops threaten food supplies and economic prosperity, potentially exacerbating broader security concerns and the global economy;
- Governmental, societal, organizational, and personal perceptions as to the legitimacy and efficacy of biological weapons can have an important impact upon the risk;
- It may not be possible to prevent all attacks; however, a coordinated series of actions can help to reduce the risk;
- A comprehensive and integrated approach is needed to prevent the full spectrum of biological threats as actions will vary in their effectiveness against specific threats.

All of this collective documentation provides an important view into how the United States both perceives the threat of a bioevent as well as the measures that have been implemented nationally to prevent such an event from occurring. Further, these documents augment robust laws that have been in existence for decades in the United States. A 1997 *Journal of American Medicine Association* article sums it up as follows:

> The US Congress has developed a comprehensive legal framework to prevent the illegitimate use of toxins and infectious agents. As part of this framework, Congress has defined as a federal crime virtually every step in the process of developing or acquiring a biological agent for use as a weapon. At the same time, Congress has vested federal law enforcement agencies with broad civil and investigative powers to enable the government to intervene before such weapons are used or even developed. Finally, Congress has directed the Centers for Disease Control and Prevention to establish a regulatory regime to monitor the location and transfer of hazardous biological agents and to insure that any use of such agents complies with appropriate biosafety requirements. [7]

Within the U.S. government, departments and agencies that engage in biodefense work are required to assess their potential research to ensure that it does not violate the BWC, ensuring that all biological experiments are for "prophylactic, protective or other peaceful purposes." For example, in the U.S. Department of Homeland Security, the body that assesses BWC compliance is called the Compliance Review Group (CRG). The CRG meets quarterly to consider all future biodefense research efforts. The potential projects are ranked according to the type of work to be conducted, the pathogens to be studied, and the quantities to be used. They are placed in one of three categories: (1) those that do not raise any compliance concern; (2) projects that might raise a perception of a compliance issue but do not involve the NSABB experiments of concern; and (3) projects that raise the perception of a compliance concern and involve one of the NSABB experiments of concern. [8]

Based on concerns about the publication of two H5N1 studies conducted in 2011 that were attempting to better understand the genetic mutations that would make the H5N1 virus both virulent and highly transmissible, the U.S. government has established a new requirement called the Dual Use Research of Concern (DURC) policy. This policy requires that all government-funded research be compared to a list of pathogens and the NSABB seven experiments of concern to ensure that those experiments that involved a pathogen on the list and at least one of the experiments of concern are listed as DURC and receive special consideration. Questions such as whether the experiment can be done differently so as not to be DURC need to be examined. Additionally, special scrutiny must be given prior to any publication of material from a DURC experiment. This sort of national implementation is clearly not the norm, as evidenced by the VERTIC statistics quoted previously and provided in Appendix G, which indicates that in most of the categories considered the global totals are around 33 percent.

National implementation should also consist of internal assessments of the degree to which a nation believed that it was in compliance with those articles of the Convention that required national attention. An assessment such as this could only be made based on internal controls such as periodic inspections required of high-containment facilities. In the United States, annual inspections are conducted for laboratories where work with especially dangerous pathogens or select agents are being used. In this way, a nation would inspect and report on critical potential "dual-use" facilities. This national-level inspection process could be augmented with monitoring bodies such as is done for the NPT with the IAEA and the CWC with the OPCW. For biodefense, this would likely include representatives from the WHO, OIE, and FAO. In this way, one could minimize the politics of the inspection that ensues when national representatives conduct on-site visits and report directly back to capitals.

Turning to the difficult question of an international verification protocol, history indicates both the sensitivity of this issue and the importance many nations attach to developing such a regime. However, several factors combine to make international verification problematic.

The accounts of previous incidents of violations of the BWC, using the Soviet Union and Iraq as examples, strongly indicate that a standard verification mechanism such as those developed for nuclear weapons cannot be directed transplanted into the biological arena. Nuclear verification is based on definitions, counting rules, transparency measures, and on-site inspections, both for monitoring destruction and counting weapons holdings. Only a limited number of facilities are able to be used for the development, assembly, and storage of nuclear weapons. The precursor material is not commonly found in the environment, but it must be mined in large quantities and processed to make fissile material suitable for use in a nuclear weapon. The equipment required for producing nuclear material is also highly specialized. This limits the ability to procure and develop new weapon material.

In contrast, biological agents are naturally occurring, and one only needs to acquire a suitable pathogen and strain in order to grow more of the biological material. The equipment used in all aspects of biological weapons development is very common; for a modest amount, one can buy the necessary equipment at a well-stocked home improvement store. For higher-end production, the more specialized equipment associated with industrial capacities can also be procured relatively easily. Many of the necessary components would be available online if one is buying it used or could be procured through supply centers that would specialize in fermentation activities such as a microbrewery. The nearly 1,400 BSL-3 and -4 facilities in the United States demonstrate the three orders of magnitude difference in the biological and nuclear verification requirements. Even if one were to triage the facilities to determine only those with suitable space to house a program of concern, the numbers would be well into the hundreds given the minimal requirements for a biological weapons facility.

If one were to inspect each of the facilities, would there be increased confidence in compliance with the BWC even if nothing out of the ordinary were found? The answer is, likely not. It would simply demonstrate no offensive biological capabilities were identified at these particular sites.

An issue that underlies the entire verification/compliance issue is the development of definitions and reporting requirements that serve as the foundation for all other elements of a potential protocol. Of course, definitions are required. But there is also great concern with strict definitions as there is the fear that they can be circumvented. Would we define the pathogens of interest in the Convention to be a certain genus, species, and serotype for a bacteria and virus and strain for a virus? What if a chimera—the melding of two different species or viruses—were to be developed? Would it escape the definition and thus be allowable or at least not covered by the Convention?

How about if the BWC definition of a potential offensive biological weapons facility were to focus on facilities with fixed fermentation vessels with more than a one-hundred-liter capacity? Would we then begin to see 99.9 liter fermenters on wheels? Obviously, this would not be a desirable outcome.

Our experience with the article-by-article review at the RevCons actually indicates that nations believe that the BWC continues to be adequate for limiting offensive biological efforts, even given such advances as genetic engineering and *de novo* synthesis of viruses and organisms. Would the strict delineation of definitions change our subjective approach to the establishment of unacceptable behavior? One would hope not, but it is and should be a concern as we would begin to establish the strict definitions that would be required for a verification regime.

The discussion of definitions and reporting ultimately leads us back to the question of the CBMs, in particular the low submission rate for these voluntary measures. It would not be possible to have a verification protocol without mandatory reporting requirements. Therefore, all member nations would need to commit to annual submissions. In making this statement, we must also begin to understand the rationale for the low rates. For this, there is no data, but rather only informal discussions with national representatives that provide some insight. Some nations believe that the CBMs are only for those nations that had offensive BW programs in the past. Some have difficulties with the forms as they are complex, written in a limited number of languages, and are certainly not intuitive. Others do not desire to provide voluntary information for secrecy reasons. Still others do not have the information readily available. For whatever reason that a nation has not submitted annual input, this would need to be corrected in order to have a viable verification protocol. Therefore, if the goal is to eventually have a BWC verification protocol, this underreporting on the CBMs must begin to change immediately.

The effort to revise the CBMs is also important to the development of a verification protocol. In an ideal world, one would desire to make changes now to the CBMs, which would begin to move closer to the data submissions that would be required under a verification protocol. To do this would require a serious discussion of the biological threats and risks that are perceived by the member nations. Once defined, they would begin to shape the reporting requirements and ultimately inform the verification protocol that would be developed.

As we are contemplating these data and reporting issues, we must also remain cognizant that the very nature of the BWC would change from a strategic, intent-based treaty to a more "standard," objective treaty. We should consider carefully what this change would mean to the unequivocal norm against the use of BW that has been established and largely held with the exception of several known bioterrorist incidents. This is not necessarily undesirable, but we must recognize that the strength of the BWC has been the unequivocal norm against the use of biological weapons as embodied in Article I.

Finally, we must agree internationally what "effective verification" entails. In the spirit of Paul Nitze, we must ensure that perfect does not become the enemy of good enough. Certainly, we would want to be able to assess whether a militarily (or strategically) significant violation has occurred. But how would we be able to judge noncompliance? As history demonstrates, assessing compliance is more art than science.

On the related issue of assessing and reporting noncompliance, the Convention does have provisions that allow for bilateral discussions and investigation using the secretary general's mechanism. The Convention provides the necessary mechanisms, yet it lacks the body with the appropriate size, expertise and "neutrality" of those that support the NPT and CWC. More on this will be discussed below.

Keeping the BWC Relevant: Measuring the Impact of Technology

Discussions about the potential for the BWC to drift in irrelevance should be cause for great concern. At the very time when biotechnology and related fields are moving at an exponential pace, the BWC is plodding along with annual experts meetings, some modest side events such as the one hosted by the United States in July 2012 that entailed a visit by several ambassadors to two U.S. government biodefense facilities, and every-five-year meetings that allow for decision making.

The structure of the BWC was set during the initial BWC document that entered into force in 1975. It called for Review Conferences held at five-year intervals. Only minor changes to the organization and structure have been taken since. After the efforts to negotiate a verification protocol collapsed in 2001, the states parties agreed to a series of annual, intersessional Meetings of Experts and Meetings of States Parties on specific themes. The success of the first series led the Sixth Review Conference to agree to a second series from 2007 to 2010. This trend was continued after the 2011 RevCon, which established an intersessional process in 2012 to 2016.

The ISU lacks the mandate, expertise, or depth to begin to keep pace with the changes that are occurring in biotechnology. This has largely been by design, because with no verification protocol, there is little perceived justification for a larger ISU. The ISU consists of three full-time personnel. With the number of conferences they are required to attend, the discussions with nations about CBM submissions, and the preparations for the annual meetings, staying abreast of changes in the biotechnology field is virtually impossible.

The role of the ISU was clearly established in the language of the Second RevCon in 1986—"administrative support and several tasks related to the confidence building measures annual submissions." Nothing more was required or desired from the element. However, today a serious discussion is needed to examine the missions of the ISU. What do we want the ISU to be able to do in terms of interface with other international elements in the biodefense community such as the WHO, FAO, OIE, and INTERPOL, to name a few? What would the member states parties expect from the ISU should there be a global pandemic caused by a bioattack?

The ISU contains no real analytical capability even for review of the CBMs. Consider the lack of expertise in the sheer number of BWC-related areas. The necessary skill sets to round out a more robust ISU would undoubtedly include a statistician for the annual submissions. It would also include human, animal, and plant disease specialists, epidemiologists, infectious disease specialists, and scientists who understand the bleeding edge of biotechnology.

In contrast, the IAEA has a full-time staff of 2,300 personnel, with a mission of "the world's center of cooperation in the nuclear field."[9] The organization has six directorates, two of which—Nuclear Safety and Security and Safeguards—have direct applicability to minimizing or eliminating the risk of proliferation of weapons-grade fissile material. The OPCW has a full-time staff of five hundred, an annual budget of 75 million Euros, and regular meetings and dialogues in The Hague. Its website lists the OPCW's tasks as having a "Technical Secretariat of the OPCW [which] is responsible for the day-to-day administration and implementation of the Convention, including inspections, while the Executive Council and the Conference of the States Parties are decision-making organs designed primarily to determine questions of policy and resolve matters arising between the States Parties on technical issues or on interpretations of the Convention."

This is not to imply that the exact missions of the IAEA or OPCW should directly transfer to the ISU in the biological realm, but rather that in terms of scale, the BWC implementation body does not even compare. This is particularly troubling if one believes that the twenty-first century is the Age of Biotechnology.

The ISU also has a role in the implementation of Article X of the Convention, the sharing of information and cooperation on biological issues. The 2011 RevCon established an Article X database that would essentially assist in linking potential donor and recipient nations. Even adding this administrative task will likely prove challenging with such a small staff.

While the question of the size and mandate is one issue that relates to long-term relevance of the BWC, another is the technical assessment of the relevance of the Convention over time based on changes in technology. While this was alluded to above, the issue warrants far greater scrutiny. Given the structural impediments and the composition of the BWC at the RevCon, making these sorts of judgments in a compressed time frame becomes problematic.

While national delegations undoubtedly do some preparation prior to coming to a RevCon and the annual experts meetings allow for the exchange of information on technical topics, it does not facilitate a global assessment of how technology will or could change the applicability of the BWC. A more formal discussion is needed to make such important assessments.

Experts must also contribute to the development of any verification protocol. Technical experts' inputs are essential to understanding the types of data and requirements that would be relevant for a mandatory data exchange that accompanies a verification protocol.

Finally, a fundamental question confronting the BWC is the natural convergence with the CWC. The issues—BW and CW—were separated in the late 1960s debate. This logical separation made sense given the status of chemical versus biological capabilities at the time. It has continued to make sense as the CWC has been largely focused on the elimination of CW stockpiles. However, once these stockpiles have been eliminated and given the emerging capabilities to synthesize biological material with chemical means, the distinction is rapidly losing relevance. What does this portend for the two conventions? Should they be logically combined as this convergence continues?

This convergence will likely also place more strain on calls for a verification protocol for the BWC. One can see the argument emerging that if the fields of biology and chemistry are converging, and the CWC has a verification protocol, then why is it not possible for a similar agreement to be reached for the BWC (under the assumption that the two conventions remain separate)?

Sharing the Benefits of Technology

A consistent complaint made by the developing nations is the lack of sharing of key biodefense technologies. Movement toward satisfying this complaint has been made by the inclusion of a donors and recipients list that could be used to pair offered capabilities with requirements. While this is viewed as a necessary step, others will feel that this alone does not address the spirit of the letter of Article X of the Convention. As a reminder, the article requires that nations "avoid hampering the economic or technological development" and assures member states "the right to participate in, the fullest possible exchange of equipment, materials and scientific and technological information for the use of bacteriological (biological) agents and toxins for peaceful purposes."

The crux of this continued argument will be that developed nations are not doing enough on this account. While developing nations may believe this is a valid complaint, no objective requirements have been established and nations are free to assist other nations with support as they see fit. For example, under the Nunn-Lugar Cooperative Threat Reduction (CTR) program, the United States has engaged broadly with the states of the former Soviet Union (FSU), including Russia, Ukraine, Georgia, Azerbaijan, Belarus, Uzbekistan, and Kazakhstan, and more recently expanded in non-FSU nations and regions including Africa and Southeast Asia. These efforts provide international assistance that falls into three basic categories: biological safety and security, threat agent detection and response, and cooperative biological research. In fiscal year 2010, this totaled approximately $250 million for the Department of Defense alone and clearly supported the United States' BWC Article X efforts. Other international efforts, such as the G-8 Global Partnership, have similar programs to the Nunn-Lugar initiative.

Another common complaint regarding the sharing of biodefense technology is that national export laws and regimes such as the Australia Group have prejudicial export practices that unfairly limit technology adaptation. From the perspective of the developing nations, the issue may be seen in this manner; however, the other side of the equation is that Article III of the Convention requires that states "undertake[s] not to transfer to any recipient whatsoever, directly or indirectly, and not in any way to assist, encourage, or induce any State, group of States or international organizations to manufacture or otherwise acquire any of the agents, toxins, weapons, equipment or means of delivery specified in Article I of the Convention." This results in a delicate balance between limiting proliferation and contributing to the biodefense capabilities of a state.

Operational Relevance of the BWC

When one meets BWC experts or views the inside of the Geneva United Nations Hall where the meetings are held, there is a sense of pace and timing that seems unaffected by events of the day. The agendas for meetings are set well in advance. Experts can remember the events of 2001 as though they were yesterday. The number of meetings between RevCons is known by date and topic well in advance.

Issues such as the H5N1 pandemic gain of function experiments and potential publication of the findings that were unfolding during the time of the 2011 RevCon seem to be irrelevant to the deliberations. One also gets the sense that the meetings unfold in set-piece fashion. Yet there is an important dilemma that the member state parties are facing.

The H5N1 issue provides some indication of the degree to which the BWC sees itself as involved in operational issues. One would have thought that if any issue was worthy of discussion in the BWC—given that it was a highly topical, policy-based issue—it was the H5N1 publication question. Yet this issue was specifically avoided. Perhaps the answer is self-evident, but it is worthy of consideration: Could the BWC play a role in the interface with other international organizations, nongovernmental organizations such as the bioassociations, or even national governments on issues such as the publication of potential articles of concern?

How about the role of the BWC in more operational issues, such as in the event of a natural disease outbreak or significant biological attack?

The BWC played no part in the 2011 *E. coli* outbreak in Europe. What if the outbreak had turned out to be a bioterror attack rather than a naturally occurring disease caused by the mishandling of agricultural products? The members generally agree (although some with less enthusiasm) that bioterror is an included threat.

What about a larger event? Consider a scenario such as the 2005 Atlantic Storm exercise that included the deliberate release of highly pathogenic smallpox at six international locations including Istanbul, Frankfurt, Rotterdam, Warsaw, Los Angeles, and New York between January 1 and January 4. The sites targeted were primarily major transportation hubs such as Penn Station in New York and the Frankfurt Airport in Germany; the lone exception was Istanbul in which the Grand Bazaar was targeted. At each location, between eight thousand and twenty-four thousand people were assumed to have been infected.[10]

What would the ISU do to support a response? Would a special meeting of the BWC member state parties be held? Would all response and recovery actions be handled by the WHO? What do the member states want the role of the ISU to be in a crisis? How about forensics and attribution? Or would those duties all fall to INTERPOL and national and regional law enforcement?

To date, no provisions have been made to engage the BWC in any operational activities or coordination. We certainly do not want to build robust mechanisms and teams to conduct response and recovery across the globe. That would be structure best left up to the individual member states or regional organizations. However, it does seem reasonable that some degree of coordination by the ISU would be a useful activity. Certainly, the WHO would be very involved in biosurveillance activities and assessing the public health impact of the disease, both immediately and into the future. The WHO would also assist with information or guidance to national governments. So what should be the role of the BWC in operational events?

Before leaving this topic, the author must comment that the suggestion is not that the BWC develop all of these capabilities, but rather that these topics have not received the discussion necessary to understand what the role of the BWC is in major policy and operational events. This should at least be a topic of debate at a future BWC event.

Managing Change within the BWC

While it is possible to control what goes on inside the conference hall in Geneva, the pace of biotechnology cannot be controlled. Over time, the effectiveness of the BWC will erode if nothing is done to ensure relevance to the events of the day. Yet one gets the sense that change is an unwelcomed intruder.

The same debates have been consistent since the entry into force of the Convention. The first among equals in this regard is verification. This issue will require great deft to manage as events of 2001 demonstrated. However, we must also realize that the cacophony will continue even if some of the nations raising concerns are acting disingenuously. In this regard, the verification debate is broken into several camps: those nations favoring a verification regime, those that are somewhat indifferent to the debate, and those that are opposed.

The United States has clearly signaled firm opposition to a return to the discussion of the verification protocol. However, if one reads between the lines, the United States has also demonstrated a new sense of transparency and openness in dealing within the Convention. Its annual CBM submission is thorough and far more comprehensive than other nations'. The CBMs have also been published on the open as well as internal ISU website. During the 2011 RevCon, Secretary of State Clinton announced a transparency visit for ambassador-level personnel to a U.S. biodefense facility. That turned into two facilities at Fort Detrick, including the Department of Defense's USAMRIID and the Department of Homeland Security's NBACC.

This is not to say that a verification protocol is on the close horizon, but rather some progress has been made. Given this progress, what would be the most appropriate manner for moving forward?

First, a BWC verification regime should be developed and implemented incrementally. The idea of a comprehensive document that establishes a full verification protocol is likely a bridge too far. The VEREX process developed a list of twenty-one measures that could be incorporated into a verification protocol. Attempting to go down the list and incorporate something from each of the categories would not be productive. Rather, a rational process to identify the risks, threats, and issues of the most concern should be developed. Based on this assessment, measures could be developed that would address major concerns.

Developing a verification mechanism will be a slow and deliberate process. It must proceed with the understanding that the foundations for verification already are largely contained within the BWC. Many of the off-site activities such as surveillance of publications and legislation, information sharing, and declarations are largely in place with the CBMs. With the U.S. unilateral transparency visit in 2012, the door has been opened for this type of on-site activity to a certain extent.

However, attempting to overlay an intrusive and comprehensive on-site component to a BWC verification mechanism would be an expensive and unnecessary outcome. Certainly, attempting to visit the 1,400 or so sites in the United States that have BSL-3 or -4 capabilities would not be reasonable. So what could a BWC verification regime reasonably entail?

In looking to a potential future verification discussion, the experiences of the CWC and the OPCW are highly instructive. The CWC has similar scale and industry issues with large numbers of civilian industrial sites. The CWC verification system is based on the monitoring of compliance on two levels, national and international. The international system is composed of three elements—declarations, routine inspections, and challenge inspections. All states parties are required to make detailed declarations, providing information on chemical weapons, chemical weapons storage facilities (CWSFs), chemical weapons destruction facilities (CWDFs), chemical weapons production facilities (CWPFs), and facilities used in the past for the development of chemical weapons. The national-level elements are much less clearly defined and include implementing legislation, data collection, and the National Authority.[11] However, in practice, the on-site work to date has focused on destruction monitoring more so than inspections of industrial sites.

In 2010, the OPCW provided funding for 125 other chemical production facilities (OCPF) inspections. One account of the CWC on-site inspection regime suggests the difficulty with these visits:

The CWC verification system is based on three "schedules," or lists of toxic chemicals and their precursors that have been developed or manufactured in the past for military purposes. The OCPFs are multipurpose chemical-production facilities that are not monitored with the same intensity as facilities that produce agents listed in the three schedules. There are some 5,000 OCPFs around the world, far more than the number of facilities that are associated with production of agents listed on the three schedules. Many of the OCPFs are in developing countries; China and India have more than 2,000 OCPFs between them. Under current rules, the Organization for the Prohibition of Chemical Weapons (OPCW), the body responsible for implementing the CWC, can inspect only a small fraction of the OCPFs each year. [12,13]

With the large number of facilities and the relatively few number of on-site inspections per year, really understanding what activities are being conducted internationally is not possible. A crosswalk of the data submission with the facts on the ground will not likely lead to any greater confidence in compliance. Unless one were to find a munitions filling station located across the street from the chemical (or in our case a biological) facility, making a declaration of a violation would be dicey. Furthermore, if one considers the experiences of UNSCOM and later United Nations Monitoring, Verification and Inspection Commission (UNMOVIC) in Iraq and the difficulty even after years of very intrusive inspections of having the confidence of declaring absolutely that a violation had occurred would remain problematic. [14]

The difficulty for the verification debate is less about the measures to be adopted in any protocol that would be established and more about the use of the information obtained through these measures. The Oxford English Dictionary defines verification as "the action of demonstrating or proving to be true or legitimate by means of evidence or testimony." Based on this standard, how could a member state party disprove a negative? Let's say that all 1,400 facilities in the United States were inspected, would that be enough to declare us "bioweapons free"? Given the small footprint required for a biological facility and the lack of signatures emitted from an offensive biological facility, even inspecting all declared facilities would only allow for a finding that no offensive biological capabilities were undertaken at the inspected facilities.

In the end, we must find a way to get to "effective verification" rather than "perfect verification." The question of what constitutes effective verification is a topic for discussion by the member states parties. However, it cannot begin with a return to the failed verification protocol, but rather it must originate at the political level with agreement concerning the strategic goals of such a mechanism. This higher-level discussion must include the consideration of the threats, risks, and issues.

The BWC CBMs discussion held throughout the 2011 review conference that attempted to clarify and refine annual national submissions was an important step; however, it did not result in a tangible outcome. Instead, this difficult work was deferred to the intersessional process. These efforts must continue with an eye toward reporting in areas where perceived threats, risks, and issues have been identified.

After more than two decades of CBM reporting and lackluster participation in the "voluntary" submission, the BWC member state parties must make the political commitment to have all nations, as part of their national obligations to the Convention, submit CBMs. Furthermore, the quality of the submissions must be greatly increased. The U.S. submission is more than 250 pages, while many other submissions from developed nations are far less comprehensive, in one case as little as ten pages.

One of the major issues confronting the BWC for the future will be the organization and structure of the Convention. The concern with the frequency and content of the meetings must be addressed. Informational meetings are held on an annual basis for experts and side events augment formal BWC meetings; however, decisional meetings are held only every five years. This is simply not adequate to address the weighty issues that the BWC must address.

The intersessional process has been an important start in expanding the agenda, but certainly not enough. An initiative during the 2011 review conference attempted to have limited decision-making authority for a defined set of issues delegated to the annual meetings, but this measure was not accepted. Regardless of the exact mechanism selected, the BWC must have more frequent opportunities for decision making. This does not necessarily imply meetings; it could entail considering written products and gaining approval within capitals.

Another important question related to the issue of meetings is attendance. The custom has been that the national delegations have been composed largely of representatives of the Conference on Disarmament (CD) in Geneva. This has been a convenient approach given the relatively light requirements associated with the CD. However, if one were to postulate a more aggressive schedule for either the CD or the BWC with more frequent meetings, it follows that delegations with greater expertise on biological issues would be beneficial.

Of course, none of this enhanced activity will be possible without some changes to the structure and mandate of the ISU. The three-person cell does superb work, yet it has virtually no ability to conduct analyses or coordinate major activities. This is both by mandate and structure. The discussion above concerning the CWC and OPCW demonstrates a far different level of activity.

The criticality of the biodefense issues of the day necessitates a more robust structure for a more relevant BWC. In this regard, the first step is to commit to changing the mandate of the ISU to include a greater analytical capability. The member states must also consider in a deliberate manner what the expectation of the ISU would be in a major BW attack. Will the ISU be involved in any way or just be an observer? In short, the BWC must have organizational structures commensurate with the level of activity that is envisioned.

The question of the relationship to the WHO, FAO, and OIE, in particular, as well as other international organizations, must be considered. To date, these relationships have developed through the informal coordination of the personnel within these organizations and the ISU. The question is, should we formalize these relationships? If so, in what way? This comes back to the questions of mandates and missions for the ISU. It also cuts to the very issue that the member state parties face: What do we expect from the BWC?

Other important or at least potentially important constituencies for the BWC include industry and the professional bioassociations. The relationship should be one of give and take. The BWC could be assisting in informing these groups about implementation requirements and best practices associated with the Convention. This is actually done by the OPCW and the U.S. State Department on behalf of the chemical industry. On the other hand, industry and bioassociations could be informing the BWC either through the ISU or to member states parties on evolutions in the field of biotechnology that could affect the treaty.

These linkages are critical as the 2001 period demonstrated. The United States walked out of the review conference at least partially based on expressed concerns by industry, in particular the pharmaceutical industry, about their apprehension concerning on-site visits leading to industrial espionage. Since many of these same issues translate directly to the chemical industry, some insights can be gleaned from these experiences.

The overarching point is that the relationship between the BWC and industry and professional organizations must be made more formal. This can only happen effectively with a change to the ISU mandate and increase in personnel.

Changes to the BWC would not be without cost. Some costs could be financial; others could result in changes to the BWC obligations such as an intrusive verification regime that could result in other unwanted effects. For example, industrial espionage could result from on-site inspections. Additional BWC requirements could inadvertently slow the progress on biotechnology by placing limits on legitimate research. Any of these negative outcomes would not be acceptable.

The cost issue is not trivial as the 2011 review conference demonstrated. The recommendation to add two personnel to the ISU staff failed as nations, based on operating on instructions from capitals to not allow any increase in cost, voted against this measure. For twenty-six state parties this would have resulted in US$19 per year in increased assessments. [15]

CONCLUSIONS

In this globalized world and in this Age of Biotechnology, we find the BWC at a strategic crossroads. Will the Convention continue to muddle through, serving as an observer in the BW debate? Or will it evolve and adapt to meet the challenges posed by the rapid advances in biotechnology?

Many see the 2011 review conference as a lost opportunity, perhaps even an international failure. In describing the final outcome, the Bioweapons Prevention Project (BWPP), which published daily reviews of the conference, summed up the final outcome as follows:

> There was a moment during the Conference in which there was language put forward for one of the ISP agenda items that would have included a conceptual discussion about how compliance might be understood. This language would have allowed verification to have been discussed within the ISP and this language was not opposed by the US, although it was a grudging acceptance. Here was the moment at which the Review Conference could have opened up the basket of verification issues by tackling the hardest of the questions that underlie the problem—what do States Parties really have to understand about each other's activities in order to have confidence, or otherwise, in the compliance of other States Parties to the BWC? Common understandings about that question during the ISP would move the debate about compliance and verification forward. But the opportunity was lost. [16]

Perhaps this critique is overly harsh. Maybe all member nations expect from the BWC is a statement of political will, an unequivocal norm, against the use of the biological weapons. But this debate should be had overtly and not languish and be considered once every five years.

Chapter Eight

Conclusions and the Way Forward

The BWC is the most important arms control treaty of the twenty-first century and in this Age of Biotechnology.

Yet the BWC truly is at a strategic crossroad. After the last review conference, many are questioning the relevance of the BWC. Less than two-thirds of the nations party to the BWC actually attended the last review conference. Few new ideas or discussion topics were introduced. No big pronouncements were made. In a very real sense the issues of the RevCon and indeed the Convention remain as they were when the BWC entered into force in 1975. Nations seemed willing to allow the review conference to drift into the old patterns of discussion. This review conference served once again as a platform for the nonaligned nations to complain about their Article X issues and the group that became known as the PRIICs (Pakistan, Russia, India, Iran, and China) to use the BWC as a forum for complaining about the U.S. position on verification rather than as a serious forum for addressing significant biodefense issues. In short, history will undoubtedly see the 2011 review conference as a missed opportunity.

Looking beyond the RevCons and more broadly at the BWC, we also see a need for greater emphasis by the international community and the member states. While more nations are member state parties, they contribute little to the implementation of the BWC. They are not even really part of the dialogue and function more as observers than participants. National implementation is sporadic at best, and relatively few nations even submit the annual "voluntary" confidence-building measures. Few actually have laws, policies, and regulations concerning the Article I provisions of the Convention. The ISU lacks any real capacity for analysis and coordination other than for the collection of the CBMs, posting them to the website and organizing and attending conferences.

But there is more to the story. The BWC operates at the very intersection of globalization, biotechnology, and security. Globalization has made the world smaller, with a greater number of human interactions. Peoples and cultures routinely interact with each other and new environments. Some of these interactions inevitably lead to tensions and potential conflicts as well as exposing people to a broader range of threats. The sheer magnitude of the biotechnology industry mandates that it will be of growing importance to our livelihood, including our health, welfare, security, and quality of life. Biotechnology shows great promise, but as the history of humankind demonstrates virtually no technology is exempt from misuse or use for nefarious purposes.

The "Ten Reasons Why the BWC Matters" embodies the importance of the treaty. As we continue to see the expansion of biotechnology into the daily lives of average citizens, the importance of having international bodies that can ensure the appropriate use of technology will become more critical. Decisions made or not made in the BWC forum could have important positive and/or negative effects on global society as well as average citizens. Our interests lie in assuring appropriate access to the benefits of biotechnology while protecting against the misuse of technology. This is not the purview of a single entity or single nation. We are in this together with a shared stake in the outcome.

Changes in biotechnology hold great promise but also the potential for severe negative consequences. This will only be exacerbated as we see a greater convergence of biotechnology with other technical areas such as informatics, nanotechnology, food security, biodiversity, and the environment. Recent innovations have demonstrated the great potential for unlocking the mysteries of disease, understanding the immune system at increasingly granular levels, and even fundamentally altering what we know about life. The time for having these discussions is before the next bioevent occurs, not at the time it is occurring.

In this prolific period when the state-of-the-art in biotechnology is increasing at an extraordinary pace, it seems only a matter of time before advances will greatly reduce the threats posed by both natural and man-made disease. Some experts believe these advances are imminent, while others see the introduction of nanotechnology as a mainstream endeavor in the 2020 to 2030 time frame as a key delineation point. Despite the question of timing, a point of widespread agreement is the certainty that advances in biotechnology will change the way we think about disease and in some cases eliminating diseases altogether. However, as the dual-use argument reminds us, the potential for ever-increasing deadly capabilities to fall into the hands of dangerous individuals, groups or states that would seek to harm innocent people on behalf of their ideologies, perceived past injustices, religious beliefs, or national interests is expanding. Furthermore, the thresholds for developing more deadly and pervasive biological threats are being lowered at a rapid technical pace well before laws, regulations, and guidelines can be established to deal with the issues. Just as with virtually all other technologies, at some point, we are likely to see these biotech capabilities used in an attack, and as the technology matures and becomes more readily available, these attacks have the potential to become large scale. The BWC has the potential to play a significant role in mitigating these emerging threats.

To date, the BWC has for many reasons not had the visibility or emphasis that other treaties have enjoyed. The result is a Convention that operates at a plodding speed while the important issues that surround biotechnology are moving at an exponential pace. Imagine watching a movie and seeing only five minutes of the film every thirty minutes. Would you be able to follow the plot? This is the manner in which the BWC is being managed. Meeting for decisions every five years in a three-week, highly choreographed event is more theater than serious work on critical issues of the day. We can and must do better.

We see many questions that must be addressed for the BWC to avoid drifting into irrelevance. What are the threats that we are facing? What should the role of the BWC be in dealing with and addressing the issues of the day with regard to emerging infectious disease, biological accidents, bioterror incidents, or state BW events? Should the BWC have an operational role? An investigative role? Or is it enough that the BWC establishes an unequivocal norm against the use of biological weapons?

We must also address the relationship between the BWC and the other international forums such as the WHO, FAO, and OIE. In the nuclear realm the IAEA has both the mandate for assisting nations in the development of nuclear technology for peaceful purposes as well as the role of monitoring the security of dangerous nuclear material and technologies. For biological issues, the WHO clearly has the role of monitoring public health, but it does not necessarily have the mission for security beyond public health. This is where the BWC fills an important void. However, the lack of structure, mandate, and overall capacity hinders the BWC from fulfilling this role. Perhaps this is acceptable, but we must collectively agree that this gap exists and identify which international body will ultimately fill this void.

The intersessional process has been an important step to assuring that BWC parties meet more frequently, but it is not enough. As a minimum, the BWC must meet for annual decisions and establish groups to report on the key issues. This would include threats and risks, the ISU mandate and organization, verification, and Article X issues. A work plan must be established to discuss these and other biodefense issues of the day.

As stated in the introduction, it is not my intention to radically change the United States' position regarding the BWC, in particular the verification issue. Rather, my thought would be to ensure that the BWC does not drift into irrelevance by failing to address the issues that are perceived to be central to the future of the Convention. Alternatively, we must ensure that any changes to the Convention do not have second- and third-order deleterious effects. To this end, we must embrace the BWC as a different type of arms control treaty and avoid the temptation for identifying easy answers; if there were easy answers remaining, they would have already been recognized and addressed.

The verification issue is central to the overall BWC debate. It is an issue that will not go away. However, it is also a very misunderstood subject. In the debate, we must avoid at all costs the pull of those who would suggest a framework that is similar to those used in nuclear arms control. The differences are fundamental. The fields are very different with material being the center of gravity for nuclear discussions and intent being the center of gravity for biological issues. Nuclear material must be processed using a highly industrialized process, which requires a large footprint. It is the exclusive purview of states. Developing an offensive biological capability requires virtually no specialized equipment either for development or deployment of a biological weapon.

The CWC and Organization for the Prohibition of Chemical Weapons (OPCW) seem more appropriate places to start; however, differences do exist between the chemical and biological domains. Still, the notion of verification or compliance being composed of national and international measures is a common theme that can be capitalized on in future discussions of BWC compliance.

Inherent in the verification debate is what constitutes "effective verification" and how one knows definitely that a violation has or has not occurred. As history demonstrates, establishing the facts on the ground is problematic, even when intrusive inspection measures are undertaken. In this debate about verification, cost will undoubtedly be an issue. Intrusive inspection regimes would be costly and likely not provide a much greater degree of assurance of compliance. Additionally, one can imagine the potential for protracted debate following an inspection if "anomalies" were identified. Are they simple technical violations of the BWC, significant cheating or noncompliance, or simply areas of dual use? These kinds of discussions and potential accusations would do little to increase openness and transparency and could even adversely affect the Convention.

In some regards, the U.S. position on verification is perplexing to the international community. It is especially confusing in light of the 2011 announcement at the review conference by Secretary of State Clinton of the U.S. intent to have a transparency visit to a biodefense facility. If the United States is willing to allow such a visit, why is a verification protocol out of the question? This is a position that the United States must spend more time articulating for the international community.

In this journey, it has not been my intent to take sides on the BWC issues confronting the member state parties, but rather to signify the critical importance of the Convention and identify the shortfalls that are contributing to the relatively low priority of the BWC. This book raises many issues that need to be addressed in this future debate and provides insights into the important questions of the day, but it also leaves many of them as unanswered as it is my belief that answering them will require the full collaboration of the member states.

The case for the importance of the Convention is not meant to imply that the BWC should usurp the work that is ongoing in other arms control negotiations. These efforts are critical as well, such as the New START treaty that reduced nuclear holdings in the United States and Russia or the United Nations Convention on the Law of the Sea Convention (UNCLOS) that as of this writing had not been ratified and that has significant economic implications. But the BWC must receive more emphasis in this Age of Biotechnology for our very security and prosperity.

Within the framework of the emerging Age of Biotechnology and with clear forethought, we must ensure the revitalization of the BWC. Furthermore, we must move to ensure the BWC gains the prominence needed to ensure the full implementation of the fifteen articles of the Convention drafted almost forty years ago yet with relevance well into the future.

To do anything less would be irresponsible.

Appendix A: The Biological and Toxin Weapons Convention (BWC)

The States Parties to this Convention,

Determined to act with a view to achieving effective progress towards general and complete disarmament, including the prohibition and elimination of all types of weapons of mass destruction, and convinced that the prohibition of the development, production and stockpiling of chemical and bacteriological (biological) weapons and their elimination, through effective measures, will facilitate the achievement of general and complete disarmament under strict and effective international control,

Recognizing the important significance of the Protocol for the Prohibition of the Use in War of Asphyxiating, Poisonous or Other Gases, and of Bacteriological Methods of Warfare, signed at Geneva on June 17, 1925, and conscious also of the contribution which the said Protocol has already made, and continues to make, to mitigating the horrors of war,

Reaffirming their adherence to the principles and objectives of that Protocol and calling upon all States to comply strictly with them,

Recalling that the General Assembly of the United Nations has repeatedly condemned all actions contrary to the principles and objectives of the Geneva Protocol of June 17, 1925,

Desiring to contribute to the strengthening of confidence between peoples and the general improvement of the international atmosphere,

Desiring also to contribute to the realization of the purposes and principles of the Charter of the United Nations,

Convinced of the importance and urgency of eliminating from the arsenals of States, through effective measures, such dangerous weapons of mass destruction as those using chemical or bacteriological (biological) agents,

Recognizing that an agreement on the prohibition of bacteriological (biological) and toxin weapons represents a first possible step towards the achievement of agreement on effective measures also for the prohibition of the development, production and stockpiling of chemical weapons, and determined to continue negotiations to that end,

Determined, for the sake of all mankind, to exclude completely the possibility of bacteriological (biological) agents and toxins being used as weapons,

Convinced that such use would be repugnant to the conscience of mankind and that no effort should be spared to minimize this risk,

Have agreed as follows:

Article I

Each State Party to this Convention undertakes never in any circumstances to develop, produce, stockpile or otherwise acquire or retain:

(1) Microbial or other biological agents, or toxins whatever their origin or method of production, of types and in quantities that have no justification for prophylactic, protective or other peaceful purposes;

(2) Weapons, equipment or means of delivery designed to use such agents or toxins for hostile purposes or in armed conflict.

Article II

Each State Party to this Convention undertakes to destroy, or to divert to peaceful purposes, as soon as possible but not later than nine months after the entry into force of the Convention, all agents, toxins, weapons, equipment and means of delivery specified in article I of the Convention, which are in its possession or under its jurisdiction or control. In implementing the provisions of this article all necessary safety precautions shall be observed to protect populations and the environment.

Article III

Each State Party to this Convention undertakes not to transfer to any recipient whatsoever, directly or indirectly, and not in any way to assist, encourage, or induce any State, group of States or international organizations to manufacture or otherwise acquire any of the agents, toxins, weapons, equipment or means of delivery specified in article I of the Convention.

Article IV

Each State Party to this Convention shall, in accordance with its constitutional processes, take any necessary measures to prohibit and prevent the development, production, stockpiling, acquisition, or retention of the agents, toxins, weapons, equipment and means of delivery specified in article I of the Convention, within the territory of such State, under its jurisdiction or under its control anywhere.

Article V

The States Parties to this Convention undertake to consult one another and to cooperate in solving any problems which may arise in relation to the objective of, or in the application of the provisions of, the Convention. Consultation and cooperation pursuant to this article may also be undertaken through appropriate international procedures within the framework of the United Nations and in accordance with its Charter.

Article VI

(1) Any State Party to this Convention which finds that any other State Party is acting in breach of obligations deriving from the provisions of the Convention may lodge a complaint with the Security Council of the United Nations. Such a complaint should include all possible evidence confirming its validity, as well as a request for its consideration by the Security Council.

(2) Each State Party to this Convention undertakes to cooperate in carrying out any investigation which the Security Council may initiate, in accordance with the provisions of the Charter of the United Nations, on the basis of the complaint received by the Council. The Security Council shall inform the States Parties to the Convention of the results of the investigation.

Article VII

Each State Party to this Convention undertakes to provide or support assistance, in accordance with the United Nations Charter, to any Party to the Convention which so requests, if the Security Council decides that such Party has been exposed to danger as a result of violation of the Convention.

Article VIII

Nothing in this Convention shall be interpreted as in any way limiting or detracting from the obligations assumed by any State under the Protocol for the Prohibition of the Use in War of Asphyxiating, Poisonous or Other

Gases, and of Bacteriological Methods of Warfare, signed at Geneva on June 17, 1925.

Article IX

Each State Party to this Convention affirms the recognized objective of effective prohibition of chemical weapons and, to this end, undertakes to continue negotiations in good faith with a view to reaching early agreement on effective measures for the prohibition of their development, production and stockpiling and for their destruction, and on appropriate measures concerning equipment and means of delivery specifically designed for the production or use of chemical agents for weapons purposes.

Article X

(1) The States Parties to this Convention undertake to facilitate, and have the right to participate in, the fullest possible exchange of equipment, materials and scientific and technological information for the use of bacteriological (biological) agents and toxins for peaceful purposes. Parties to the Convention in a position to do so shall also cooperate in contributing individually or together with other States or international organizations to the further development and application of scientific discoveries in the field of bacteriology (biology) for prevention of disease, or for other peaceful purposes.

(2) This Convention shall be implemented in a manner designed to avoid hampering the economic or technological development of States Parties to the Convention or international cooperation in the field of peaceful bacteriological (biological) activities, including the international exchange of bacteriological (biological) agents and toxins and equipment for the processing, use or production of bacteriological (biological) agents and toxins for peaceful purposes in accordance with the provisions of the Convention.

Article XI

Any State Party may propose amendments to this Convention. Amendments shall enter into force for each State Party accepting the amendments upon their acceptance by a majority of the States Parties to the Convention and thereafter for each remaining State Party on the date of acceptance by it.

Article XII

Five years after the entry into force of this Convention, or earlier if it is requested by a majority of Parties to the Convention by submitting a proposal to this effect to the Depositary Governments, a conference of States Parties to the Convention shall be held at Geneva, Switzerland, to review the operation of the Convention, with a view to assuring that the purposes of the preamble and the provisions of the Convention, including the provisions concerning negotiations on chemical weapons, are being realized. Such review shall take into account any new scientific and technological developments relevant to the Convention.

Article XIII

(1) This Convention shall be of unlimited duration.

(2) Each State Party to this Convention shall in exercising its national sovereignty have the right to withdraw from the Convention if it decides that extraordinary events, related to the subject matter of the Convention, have jeopardized the supreme interests of its country. It shall give notice of such withdrawal to all other States Parties to the Convention and to the United Nations Security Council three months in advance. Such notice shall include a statement of the extraordinary events it regards as having jeopardized its supreme interests.

Article XIV

(1) This Convention shall be open to all States for signature. Any State which does not sign the Convention before its entry into force in accordance with paragraph (3) of this Article may accede to it at any time.

(2) This Convention shall be subject to ratification by signatory States. Instruments of ratification and instruments of accession shall be deposited with the Governments of the United States of America, the United Kingdom of Great Britain and Northern Ireland and the Union of Soviet Socialist Republics, which are hereby designated the Depositary Governments.

(3) This Convention shall enter into force after the deposit of instruments of ratification by twenty-two Governments, including the Governments designated as Depositaries of the Convention.

(4) For States whose instruments of ratification or accession are deposited subsequent to the entry into force of this Convention, it shall enter into force on the date of the deposit of their instruments of ratification or accession.

(5) The Depositary Governments shall promptly inform all signatory and acceding States of the date of each signature, the date of deposit of each instrument of ratification or of accession and the date of the entry into force of this Convention, and of the receipt of other notices.

(6) This Convention shall be registered by the Depositary Governments pursuant to Article 102 of the Charter of the United Nations.

Article XV

This Convention, the English, Russian, French, Spanish and Chinese texts of which are equally authentic, shall be deposited in the archives of the Depositary Governments. Duly certified copies of the Convention shall be transmitted by the Depositary Governments to the Governments of the signatory and acceding states.

IN WITNESS WHEREOF the undersigned, duly authorized, have signed this Convention.

DONE in triplicate, at the cities of Washington, London and Moscow, this tenth day of April, one thousand nine hundred and seventy-two.

Appendix B: Ten Reasons Why the BWC Matters

The premise of this book is that the BWC is the most important arms control treaty of the twenty-first century. This is a bold statement that requires more than simple assertion. Ten underlying rationale support the belief concerning the criticality of the BWC in this Age of Biotechnology.

The Biological and Toxin Weapons Convention (BWC) ...

1. ...eliminates an entire class of dangerous weapons.
2. ...provides an unequivocal norm against the use of biological weapons.
3. ...provides an international forum for dialogue concerning biological issues.
4. ...has an important economic dimension.
5. ...is gaining more importance as the spectrum of potential biological threats grows.
6. ...provides a forum for coordinating preparedness and response capabilities against a spectrum of biological threats.
7. ...is an arms control agreement that relates directly to public health, the environment, food security, and biodiversity.
8. ...provides direct linkages to international security mechanisms.
9. ...relates to dual-use capabilities in a way that no other arms control treaty does.
10. ...has responsibilities for implementation that runs the gambit from international organizations to the individual.

Appendix C: Confidence-Building Measures

The measures include:

Measure A, Part 1: Exchange of data on research centers and laboratories that meet very high national or international safety standards (WHO BL4/P4).

Measure A, Part 2: Exchange of information on national biological defense research and development (R&D) programs, including declarations of facilities where biological defense R&D programs are conducted. This measure also includes information relating to contractors and on available publications.

Measure B: Exchange of information on outbreaks of infectious diseases and similar occurrences caused by toxins that seem to deviate from the normal pattern.

Measure C: Encouragement of publication of results of biological research directly related to the Convention and promotion of use of knowledge.

Measure D: Active promotion of contacts between scientists, other experts, and facilities engaged in biological research directly related to the Convention, including exchanges and visits for joint research on a mutually agreed basis.

Measure E: Declaration of legislation, regulations, and other measures, including exports and/or imports of pathogenic microorganisms in accordance with the Convention.

Measure F: Declaration of past activities in offensive and/or defensive biological R&D programs since January 1, 1946.

Measure G: Declarations on vaccine production facilities, licensed by the State Party for the protection of humans.

Appendix D: Arms Control Definitions

Arms control has come to mean different things to different people, and its meaning has evolved throughout history. Arms control agreements existed prior to World War II (1939–1945); however, the modern arms control effort began in earnest after the Cuban Missile Crisis of 1962.[1] During the postwar period, the theory and practice of modern arms control came into being. The focus was on dealing with the mistrust between the United States and the Soviet Union. Arms control was seen as a way to limit holdings of weapons systems, control behavior, and increase transparency and stability.

One source provides four definitions of arms control from the 1960 to 1961 period, during which modern arms control was developed:[2]

> We believe that arms control is a promising, but still only dimly perceived, enlargement on the scope of military strategy. It rests essentially on the recognition that our military relation with potential enemies is not one of pure conflict and opposition, but involves strong elements of mutual interest in the avoidance of a war that neither side wants, in minimizing the costs and risks of the arms competition, and in curtailing the scope and violence of war in the event it occurs.
> —Thomas Schelling and Morton Halperin, *Strategy and Arms Control* (1961)
>
> The goal of responsible arms control measures must be to determine, free of sentimentality, not how to eliminate retaliatory forces but how to maintain an equilibrium between them . . . A control system will add to stability if it complicates the calculations of the attacker and facilitates those of the defender. Or, put another way, the objective should be to increase the uncertainty about the possibility of success in the mind of the aggressor and to diminish the vulnerability of the defender . . . The primary goal of any arms control

scheme must be to increase stability. A precondition is that both sides should
strive to develop invulnerable retaliatory forces.
—Henry Kissinger, "Arms Control, Inspection and Surprise Attack"

It is useful to think generally of arms control as a cooperative or multilateral
approach to armament policy—where "armament policy" includes not only the
amount and kind of weapons and forces in being, but also the development,
deployment, and utilization of such forces, whether in periods of relaxation, in
periods of tension, or in periods of shooting wars . . . The basic goal of arms
control . . . is to reduce the hazards of present armament policies by a factor
greater than the amount of risk introduced by the control measures themselves.
—Donald G. Brennan, "Setting and Goals of Arms Control"

The concept of "arms control" includes any agreement among several powers
to regulate some aspect of their military capability or potential.
—Robert Bowie, "Basic Requirements of Arms Control"

The National Security Strategy of 1990, the first NSS to be published after
the fall of the Berlin Wall and at the end of the Cold War, provides an
important data point for the role of arms control in national security:[3]

Arms control is a means, not an end; it is an important component of a broader
policy to enhance national security. We will judge arms control agreements
according to several fundamental criteria:

First, agreements must add to our security. Our objective is to reduce the
incentives, even in crisis, to initiate an attack. Thus, we seek not reductions for
reductions' sake, but agreements that will promote stability. We will work to
reduce the capabilities most suited for offensive action or preemptive strike.

Second, to enhance stability, we favor agreements that lead to greater predictability in the size, nature, and evolution of military forces. Predictability
through openness expands the traditional focus of arms control beyond just
military capabilities and addresses the fear of aggressive intent.

Third, agreements are effective only if we can verify compliance. As we
broaden our agenda to include issues like chemical and missile proliferation,
verification will become an increasingly difficult challenge, but effective verification will still be required. We want agreements that can endure.

Finally, since the security of the United States is indivisible from that of its
friends and allies, we will insist that any arms control agreements not compromise allied security.

It is interesting that one of the criteria is the ability to verify compliance. This
NSS was written around the time that the VEREX group engaged in its
efforts concerning development of a BWC verification protocol after the

questionable Soviet activities concerning Sverdlovsk and the "yellow rain" incidents and immediately following the defection of Vladimir Artemovich Pasechnik, who had served as a senior Soviet scientist within the offensive BW program.

Today, our concept of arms control has evolved again to include such activities such as nonproliferation and the Cooperative Threat Reduction (CTR) program. And instead of a focus on managing U.S.-Soviet (or Russian) relations, arms control has become universal, with treaties intending to involve the global security community. While the BWC was developed during the height of the Cold War, one could argue that it was ahead of its time in being one of the first treaties to include all nations desiring to participate and was certainly the first of the CBRN treaties to have, by necessity, broad international membership.

Of course, arms control is a two-edged sword and as such, it cuts both ways. Transparency and openness into the affairs of another nation translate into loss of security and even sovereignty. It means that while one nation might be able to gain insights into the activities of another nation, an expectation of reciprocity accrues.

Appendix E: Principles of Verification

From Review of the Implementation of the Recommendations and Decisions Adopted by the General Assembly at Its Tenth Special Session: Report of Disarmament Commission, The United Nations, July 1, 1996. http://www.securitycouncilreport.org/atf/cf/%7B65BFCF9B-6D27-4E9C-8CD3-CF6E4FF96FF9%7D/Disarm%20A51182.pdf

The Disarmament Commission considers that the following general principles elaborate upon or add to those stated in the Final Document of the Tenth Special Session of the General Assembly. While further work can be done to formulate adequately these and other principles relating to verification, the following is a nonexhaustive listing of such principles:

(1) Adequate and effective verification is an essential element of all arms limitation and disarmament agreements.

(2) Verification is not an aim in itself, but an essential element in the process of achieving arms limitation and disarmament agreements.

(3) Verification should promote the implementation of arms limitation and disarmament measures, build confidence among States and ensure that agreements are being observed by all parties.

(4) Adequate and effective verification requires employment of different techniques, such as national technical means, international technical means, and international procedures, including on-site inspections.

(5) Verification in the arms limitation and disarmament process will benefit from greater openness.

(6) Arms limitation and disarmament agreements should include explicit provisions whereby each party undertakes not to interfere with the agreed methods, procedures, and techniques of verification, when these

are operating in a manner consistent with the provisions of the agreement and generally recognized principles of international law.

(7) Arms limitation and disarmament agreements should include explicit provisions whereby each party undertakes not to use deliberate concealment measures which impede verification of compliance with the agreement.

(8) To assess the continuing adequacy and effectiveness of the verification system, an arms limitation and disarmament agreement should provide for procedures and mechanisms for review and evaluation. Where possible, time frames for such reviews should be agreed in order to facilitate this assessment.

(9) Verification arrangements should be addressed at the outset and at every stage of negotiations on specific arms limitation and disarmament agreements.

(10) All States have equal rights to participate in the process of international verification of agreements to which they are parties.

(11) Adequate and effective verification arrangements must be capable of providing, in a timely fashion, clear and convincing evidence of compliance or noncompliance. Continued confirmation of compliance is an essential ingredient to building and maintaining confidence among the parties.

(12) Determinations about the adequacy, effectiveness, and acceptability of specific methods and arrangements intended to verify compliance with the provisions of an arms limitation and disarmament agreement can only be made within the context of that agreement.

(13) Verification of compliance with the obligations imposed by an arms limitation and disarmament agreement is an activity conducted by the parties to an arms limitation and disarmament agreement or by an organization at the request and with the explicit consent of the parties, and is an expression of the sovereign right of States to enter into such arrangements.

(14) Requests for inspections or information in accordance with the provisions of an arms limitation and disarmament agreement should be considered as a normal component of the verification process. Such requests should be used only for the purposes of the determination of compliance, care being taken to avoid abuses.

(15) Verification arrangements should be implemented without discrimination, and, in accomplishing their purpose, avoid unduly interfering with the internal affairs of State parties or other States, or jeopardizing their economic, technological and social development.

(16) To be adequate and effective, a verification regime for an agreement must cover all relevant weapons, facilities, locations, installations, and activities.

Appendix F: U.S. Ambassador Donald Mahley's Statement

STATEMENT BY THE UNITED STATES TO THE AD HOC GROUP OF BIOLOGICAL WEAPONS CONVENTION STATES PARTIES
at http://usinfo.org/wf-archive/2001/010725/epf314.htm.

Geneva, Switzerland
July 25, 2001
Mr. Chairman, Colleagues:

I take the floor today after twenty-three sessions of the Ad Hoc Group, spanning some six and a half years of negotiation trying to develop a legally-binding document to enhance confidence in compliance with the Biological Weapons Convention. The relevance of our objective has not diminished over those years. Everyone should understand the importance the United States places on the Biological Weapons Convention and the global ban on biological weapons it establishes.

No nation is more committed than the United States to combating the BW threat. This is a threat we face not only at home but also abroad. Our forces and our friends and allies may well be the victims of this weapon of terror and blackmail. We must counter this complex and dangerous threat with a full range of effective instruments—nonproliferation, export controls, domestic preparedness, and counterproliferation. We are firmly committed to combat the spread of biological weapons.

After years of arduous negotiation, with the sterling work of numerous friends of the chair to facilitate discussions about specific issues, this group had gone as far as that technique would permit in resolving individual issues and questions along the model set forth in the original rolling text in 1997.

You, Mr. Chairman, then undertook the challenging and onerous task of proposing a set of mutual compromises, based on that rolling text, as a potential way to bring the negotiations to closure in a short period. The United States congratulates you on the effort you have made to resolve very contentious issues.

The United States has subjected the "Composite Text" proposal to detailed scrutiny. As every veteran of these negotiations will recall, the United States has had serious issues with both individual proposals and the general approach to some issues throughout these negotiations. Those concerns and requirements have not changed— indeed, they remain one of the consistent criteria against which the United States has evaluated this text.

In addition to the text, we have looked at the overall issue of biological weapons threat. Our approach for doing so is comprehensive, and includes new, affirmative ideas for strengthening the Biological Weapons Convention. We believe we can strengthen the Biological Weapons Convention through multilateral arrangements. To be valuable, however, we believe any approach must focus on effective innovative measures.

The review conducted in Washington has encompassed more than the substantive content of individual issues. We recognize that in any negotiation no individual country can dictate the outcome on all elements of the final text. We also recognize that the proposals in the "Composite Text" frequently do reflect values and views the United States has inserted into the negotiation. However, as indicated in the Twenty-Third session and for years before that, the United States still has very serious substantive difficulties with the textual proposals in these negotiations. We have recognized the substantive and political values many of the participants here attach to a successful completion of the Protocol, and that the demonstrated ability to achieve a consensus result, even if it cannot satisfy every party's preferred outcomes, is perceived as a potential benefit in itself. The United States view, therefore, has been considered at the most senior levels of our government. We continue to have problems and certainly are not going to mislead anyone on this point.

After extensive deliberation, the United States has concluded that the current approach to a Protocol to the Biological Weapons Convention, an approach most directly embodied in CRP.8, known as the "Composite Text," is not, in our view, capable of achieving the mandate set forth for the Ad Hoc Group, strengthening confidence in compliance with the Biological Weapons Convention. One overarching concern is the inherent difficulty of crafting a mechanism suitable to address the unique biological weapons threat. The traditional approach that has worked well for many other types of weapons is not a workable structure for biological weapons. We believe the objective of the mandate was and is important to international security, we will therefore be unable to support the current text, even with changes, as an appropriate outcome of the Ad Hoc Group efforts.

The draft Protocol will not improve our ability to verify BWC compliance. It will not enhance our confidence in compliance and will do little to deter those countries seeking to develop biological weapons. In our assessment, the draft Protocol would put national security and confidential business information at risk.

The United States intends to develop other ideas and different approaches that we believe could help to achieve our common objective of effectively strengthening the Biological Weapons Convention. We intend to explore those ideas and other alternative approaches during the next several months with the goal of reaching a consensus on a new approach for our shared objective.

There is no basis for a claim that the United States does not support multilateral instruments for dealing with weapons of mass destruction and missile threats. We strongly support the Australia Group, and will be working actively to strengthen it at its next meeting in Paris October 1–4. Indeed, we support all multilateral arms control, nonproliferation, and export control regimes that are currently in force, such as the NPT, the CWC, the BWC, MTCR, NSG, IAEA, Zangger Committee, and the Wassenaar arrangement.

Let me outline some of the reasons for reaching the conclusion I have just announced about this Protocol. As I noted earlier, many of these will not be new or surprising. They reflect positions the United States has advanced repeatedly throughout these negotiations.

Objectives

One central objective of a Protocol is to uncover illicit activity. Traditionally, this has meant seeking regular on-site inspections of locations potentially able to conduct such activity, the shorter-notice and the more intrusive, the better. Always, there is a balance between pursuing illicit actions and protecting legitimate national security and proprietary information unrelated to illicit activity.

In the draft Protocol, there is an inherent dilemma associated with the question of on-site activities. The provisions for on-site activity do not offer great promise of providing useful, accurate, and complete information to the international community. However, when we examined the prospects of the most intrusive and extensive on-site activities physically possible—which we believed were likely not acceptable to most other countries—we discovered that the results of such intrusiveness would still not provide useful, accurate, or complete information.

One objective is to agree on a declaration base that would provide reasonable inventories of activity in a country relevant to the underlying Biological

Weapons Convention. Our assessment of the range of facilities potentially relevant to the Convention indicates that they number, at least in the case of the United States, in the thousands, if not the tens of thousands. In addition, their number and locations change on an irregular but frequent basis. Thus, we had no hope that any attempt at a comprehensive declaration inventory would be accurate, timely, or enduringly comprehensive.

In short, after extensive analysis, we were forced to conclude that the mechanisms envisioned for the Protocol would not achieve their objectives, that no modification of them would allow them to achieve their objectives, and that trying to do more would simply raise the risk to legitimate United States activities.

This is not a new perspective. We have voiced it since the initial negotiating sessions in 1995. The United States has worked with other countries to try to find the way to create an appropriate balance in the draft Protocol. However, despite the efforts of many, we are forced to conclude that an appropriate balance cannot be struck that would make the draft Protocol defensible as an instrument whose utility outweighs its risk.

The Paradigm

Another key objective for a Protocol to strengthen the Biological Weapons Convention would be to deter or complicate the ability of a rogue state to conduct an illicit offensive biological weapons program. These negotiations have worked from the outset on the model of regimes that have gone before. The most frequently cited paradigm for our work has been the Chemical Weapons Convention. Indeed, many of the arguments and justifications for the scope and nature of activities envisioned under the draft Protocol have used the CWC as the example of comparison.

This is, unfortunately, seriously flawed. When developing the ban on chemical weapons, the question of dual-capability was in the forefront. It was, and is, a legitimate question with respect to the CWC, since the immediate precursors of chemical agents require production facilities capable of making chemical agents, but have legitimate commercial applications. The same kind of dual-capability issue exists in biology to an even greater extent.

In chemical manufacturing, although the precursors have legitimate application, the economics of their production dictates making them in a limited number of facilities. Such facilities, because of the toxicity and corrosiveness of the precursors, have recognizable infrastructure requirements.

When setting up the CWC, we were able to require universal declaration for such facilities, and then establish an international regime that would visit each such facility on a regular and repeated basis. If there were such a recognizable facility that were not declared, the very lack of declaration would be sufficient to raise questions about its role and function.

The Ad Hoc Group quickly recognized that no such cataloging was possible with respect to biology to biological facilities. Almost any facility that does biological work of any magnitude possesses the capability, under some parameters, of being diverted to biological weapons work. Trying to catalog them all would be tantamount to impossible. Likewise, visiting even those selected—almost arbitrarily—for declaration on the same universal and regular basis as the CWC would require an international organization of the size and possession of rare skills among its employees that no one in the Ad Hoc Group was willing to contemplate.

What we are left with, then, is a regime that contemplates—at best— declaration of an almost randomly-selected set of facilities from among those actually relevant to a potential proliferator. To compound the difficulty, among that random sample of facilities, regular on-site activity would take place at only a random sample of even that sub-set. And, given the distribution of biological activity around the world, despite the best efforts at finding a "smoothing function" to distribute on-site activity, the overwhelming bulk of such activity would take place on the territory of those States Parties least likely to be proliferation candidates.

In the considered judgment of the United States, the small scope of applications of this kind of twice-removed randomness, coupled with a required emphasis on the wrong targets from among the susceptible population, simply does not provide anything remotely resembling a deterrent function on a proliferator, even a non-state actor. We therefore conclude that the conceptual approach used in the current negotiating effort fails to address the objective we have sought throughout the negotiations. This approach, although relevant in the references to non-biological areas used throughout the negotiations, simply does not apply to biology. If we are to find an appropriate solution to the problem, we need to think "outside the box." It will require new and innovative paradigms to deal with the magnitude of biological activity that can be a threat, the explosively changing technology in the biological fields, and the varied potential objectives of a biological weapons program. We simply cannot try to patch or modify the models we have used elsewhere.

Biodefense Issues

Defense against biological weapons is of great concern to the United States. As we have stated repeatedly, any Protocol needed to ensure that the ability to protect against those who would violate the norm of abolishing biological weapons was not impaired. The United States has the most extensive biodefense program described by any participant in these negotiations. The United States therefore has more national security equities directly at risk through this Protocol than any other participant. At the same time, the potential downside of undercutting biodefense efforts is not limited to the United

States. We share the results of our efforts with other countries in assisting them to protect themselves against potential biological weapons attack. Our concerns, then, are not limited to self-protection. They are concerns that should be relevant to many of the countries in this room.

We recognize that finding a balance of protection and disclosure has been especially difficult in the biodefense arena. The proposal in the "Composite Text" is far from what some countries have suggested, and even incorporates a number of elements the United States has demanded. However, there are still provisions in the current proposals we believe would be inimical to legitimate national security efforts.

More importantly, as we have analyzed the options, we came to the conclusion that the same inherent flaw I described earlier is present in the approach to biodefense. Between declarations on biodefense and other categories, such as working with listed agents, the current proposals do not provide sufficient protection. At the same time, the exclusions in declarations would permit a potential proliferator to conceal significant efforts in legitimately undeclared facilities. Conversely, if we try to make the declarations comprehensive enough to capture all biodefense activity, the level of risk to legitimate and sensitive national security information becomes truly unbearable.

On-Site Activity Utility

Earlier I noted the dysfunction of concentrating on-site activity in places that would be largely irrelevant to possible biological weapons concerns. This alone detracts seriously from any value for the objectives of the Ad Hoc Group. However, there is a second liability of on-site activity as envisioned in the "Composite Text."

The activities outlined to take place on a regular basis, transparency visits, actually risk damage to innocent declared facilities, despite the fact that they would have almost no chance of discovering anything useful to the BWC if they took place at a less-than-innocent facility. This risk is a two-edged sword: proprietary or national security information may be at risk, and/or the activity may serve to misdirect world attention into non-productive channels.

A number of safeguards have been inserted into procedures to protect information not relevant to the BWC. Those safeguards are insufficient to eliminate unacceptable risks to proprietary or national security information. The nature of proprietary information in the biological field is very diverse. It ranges from overall capacities, which reveal market size and profit potential, to routine physical production configurations that provide efficiency and output advantages. Protecting such a diverse and innocent-seeming range of information would require facilities to exercise the protections incorporated in the draft Protocol language extensively and, even then, they would have no firm assurance proprietary information could not be inferred from what was seen by inspectors.

At the same time, the very exercise of the protections incorporated in the draft language could misdirect the attentions of the international community. Countries, or competitors, with economic or political agendas of disruption could raise unfounded allegations. Such allegations would be refutable only with economic or national security costs, and refutation after the fact would likely already have resulted in commercial damage to private firms. The United States, with its visibility in the world, is perhaps more sensitive to such a situation than some. Our concern, however, is not for the United States alone. We simply cannot agree to make ourselves and other countries subject to such risks when we can find no corresponding benefit in impeding proliferation efforts around the globe.

Constitutional and Ratification Issues

Throughout this negotiation, we have made all our colleagues aware of the constraints we face in achieving ratification of an international agreement. This is, to the United States, a crucial component of any outcome. We do not believe in negotiating, nor signing, agreements that do not support their stated objectives. At the same time, the United States operates in a specific Constitutional framework. It requires the executive branch of our government, to submit the results of any negotiation to the United States Senate for advice and consent to ratification. In good faith throughout these deliberations, we have brought to the attention of the Ad Hoc Group the issues where we believed there were explicit requirements to allow the United States to achieve ratification. We have operated on the assumption—which we still believe is valid—that creating an instrument that would preclude United States participation was not in the best interests of this negotiation or of the Biological Weapons Convention.

We also have explicitly recognized that some of the conditions necessary to satisfy these requirements would theoretically allow abuse. For our own part, the United States does not use such devices in an abusive fashion. We also believe that if others were to do so, the nature of their use would be obvious to any objective observer, and the international community could draw appropriate conclusions. Thus, we do not believe the potential abuse argument outweighs our own responsibility to create an instrument to which we believe we could become an active party.

There are elements of the "Composite Text" draft that violate the requirements with respect to this issue. I do not intend to try to detail them today—they are familiar to all in this room who have sat through the numerous Ad Hoc Group sessions where we have detailed both the appropriate solution and the rationale for it.

However, the result is a text that, even if the United States were con-vinced had substantive merit in achieving its stated objectives to strengthen the Biological Weapons Convention, would not be one we could predict with reasonable probability the United States would become a party to. This is, in our view, a futile effort. We deeply regret if the nature of these specific requirements was not made clear throughout the negotiations, but it was certainly not for lack of effort on the part of the United States delegation.

Export Controls

The nations involved in this negotiation should be commended for their ability to attack a subject with as many divergent national views such as global control of biotechnology. I have referred several times to the central objective of the United States in undertaking this work: to assist in the global effort to stem, or at least inhibit, proliferation of biological weapons. We believe that remains an essential goal for international security.

Some of the participants in this negotiation, however, have approached the situation with a different mix of national priorities. They view the issue as much from the perspective of technological development as from direct se-curity enhancement measures. While the United States agrees with the con-cept that global technological development in biotechnology helps create a more secure environment, we view this as a subordinate element to the compliance-enhancement aspects of any Protocol to the Biological Weapons Convention.

The Convention is, after all, a disarmament treaty, not a trade treaty. There are competent organizations throughout the world whose principal function is to fight disease, enhance trade, and promote development. The United States supports those organizations, and applauds their successes in their own areas of competence.

Other delegations appear to disagree fundamentally with our assessment. Just this week, we have heard that "Confidential Proprietary Information (CPI) is the concern of only a few advanced countries, where National Secur-ity Information is the concerns [*sic*] of all States Parties. . . . (T)herefore, my delegation expects the deletion of the references to CPI in the final text." In addition, we hear that "any parallel export control regimes have to be dis-solved after the Protocol enters into force (for States Parties to the future Protocol)."

We have explained at length why ignoring the protection of legitimately sensitive information, both for proprietary and national security reasons, is an essential element of focusing any instrument on the disarmament objectives we should be seeking rather than trade enablements, which should be the purview of other organizations. We also take seriously the threat of biologi-cal weapons proliferation. A Protocol should be, if it were properly focused and implemented, another instrument in the set of tools countering proliferation of weapons of mass destruction. Never has the argument been made successfully that it could have become the single answer to the proliferation problem. To insist that other effective tools be forfeited in order to establish a Protocol is an indication of the wide gap between demands and possible solutions still existing in these negotiations.

We do not believe the Ad Hoc Group product, or the international organization—affectionately known as the OPBW—envisioned by the current draft Protocol, is an appropriate substitute for those other organizations. In fact, we fear that the inevitable competition of alternative international organizations with overlapping mandates could actually impede some of the effectiveness of those other already existing organizations.

Likewise, we are perplexed by the arguments of some participants in these negotiations that commitments in the areas of trade and development are necessary "prices" to pay for the security-enhancing compliance measures envisioned in other elements of the draft Protocol. Global political situations would indicate that the very countries trying hardest to argue for compensation to agree to security enhancement are those most likely to have a biological weapons threat to their own security. The logic of their position is not apparent.

From the beginning, the Protocol the Ad Hoc Group has sought has been an additional tool to address the biological weapons threat. We all recognize that the threat is both real and growing. Other efforts already exist to address the threat, including the BWC itself. While they have not eliminated the problem, they nonetheless have been useful in retarding the threat.

The United States believes very strongly in employing all available means to enhance international security. One of the things we will not allow is any degradation of those tools we already have to fight a serious challenge to security. Throughout these negotiations, some participants have attempted to do just that. Such an effort is flatly unacceptable to the United States.

It is the responsibility of all of us, since we are already parties to the Biological Weapons Convention, to inhibit or prevent biological weapons being in the hands of any state or party whatsoever, by both national and international means. We take that national responsibility very seriously. To the degree we can enhance our efforts through cooperation with other states parties, we will continue to do so. Efforts to constrain, impede, or eliminate such efforts will be unacceptable to the United States now and at any time in the foreseeable future. Those who think there is any flexibility on this point in the United States are sadly mistaken, and should abandon any such pursuits.

Disturbing Negotiating Positions

Some participants in these negotiations have also sought outcomes that are, frankly, disturbing to the United States. We do not understand, even after repeated explanations, the rationale for such efforts. We can only urge that states reexamine their basis for such positions.

The mandate of the Ad Hoc Group clearly states that any Protocol must not abridge, diminish, or otherwise weaken the Biological Weapons Convention. The United States has tried to keep that principle in mind whenever we have proposed measures or other elements for the draft Protocol.

We must wonder, though, when we are asked to consider provisions that would constrict the potential scope of the prohibitions in the Convention by fixing the meaning of terms in the Convention itself. We have heard repeatedly about the flexibility needed to keep up with explosively changing technology. It seems to us that efforts in contrary directions cannot be in the interests of the object and purpose of the Convention.

Likewise, we have long held that seeing the actual effects of a biological weapons program would be one of the less ambiguous issues in evaluating potential threat. While less ambiguous, such effects are not unambiguous. It therefore seemed to us that being able to examine such effects, including disease outbreaks, was an important capability for any Protocol regime. Attempts to restrict such investigations do not seem in the best interests of all parties.

CONCLUSION

I apologize for the length of this intervention. As I stated at the outset, the United States understands and appreciates the amount of effort, and the amount of compromise, that have marked the negotiations to this point. We agree with the assessment that it was time to move from the rolling text to a composite text in an effort to formulate compromise solutions to outstanding issues. We have analyzed those efforts from both a political and substantive perspective, recognizing the sincere desire of most of the participants to reach an outcome that would have a product ready for consideration and signature by States Parties.

The United States does believe that many, if not all, of the difficulties I have outlined today are things the participants in this room have heard, repeatedly, over the last six years. These are not new ideas the United States has just now formulated—they are long-standing concerns. At the same time, we recognize that no country in a multilateral negotiation achieves all of its desired positions, and that some of the compromises reflected in the composite text are difficult for others to accept.

Others in this room have the same objectives as the United States for a Protocol—enhancing international security. The various expressions of support for the composite text we have heard clearly indicate that others have evaluated the draft Protocol and have concluded that, however imperfect, it does satisfy those objectives. Regrettably, the United States has come to a different conclusion.

We have spent the effort to examine the text in detail, and at the senior-most levels. We have also examined the principles on which the text is based. We have looked for a set of specific changes that could alter our fundamental conclusions. These intensive reviews have led us to conclude that this effort simply does not yield an outcome to which we would be prepared to agree. I have outlined above some of the reasons why we have reached this conclusion. Because the difficulties with this text are both serious and, in many cases inherent in the very approach used in the text, more drafting and modification of this text would, in our view, still not yield a result we could accept.

Some have argued both publicly and privately that not having this Protocol will weaken the BWC itself. The United States categorically rejects that supposition. Let me re-emphasize that the U.S. fully supports the global ban on biological weapons embodied in the BWC, and remains committed to finding effective ways to strengthen the overall regime against the BW threat, including multilateral ones. The United States will, therefore, work hard to improve—not lessen—global efforts to counter both the BW threat and the potential impact such weapons could have on civilization. And we would reply to those who cry that not having this Protocol weakens the global norm against BW that there absolutely is no reason that kind of reaction need occur. It will happen only if we convince ourselves that it is happening, and we would urge others to join with us in ensuring such a reaction does not take place.

Thank you Mr. Chairman and colleagues.

Appendix G: National Implementation

National Implementation of the BWC: Statistics Based on VERTIC's Legislation Surveys (as at March 23, 2011).[1]

Measures	Africa: Number (%) of 45	Asia: Number (%) of 36	Eastern Europe: Number (%) of 15	Latin America and the Caribbean: Number (%) of 31	Western Europe and Other States: Number (%) of 9	Global: Number (%) of 136
DEFINITIONS						
Biological weapon	1 (2%)	5 (14%)	1 (7%)	3 (10%)	3 (33%)	13 (10%)
CRIMES						
Develop biological weapons and penalties	5 (11%)	5 (14%)	8 (53%)	6 (19%)	7 (78%)	31 (23%)
Manufacture/produce biological weapons and penalties	6 (13%)	11 (31%)	13 (87%)	11 (35%)	7 (78%)	48 (35%)
Acquire biological weapons and penalties	5 (11%)	9 (25%)	11 (73%)	7 (23%)	6 (67%)	38 (28%)
Stockpile/store biological weapons and penalties	3 (7%)	5 (14%)	9 (60%)	7 (23%)	8 (89%)	32 (24%)
Possess/retain biological weapons and penalties	6 (13%)	7 (19%)	8 (53%)	13 (42%)	6 (67%)	40 (29%)
Transfer biological weapons and penalties	3 (7%)	10 (28%)	12 (80%)	12 (39%)	6 (67%)	43 (32%)
Use biological weapons and penalties	6 (13%)	10 (28%)	11 (73%)	12 (39%)	4 (44%)	43 (32%)
Engage in activities involving dangerous biological agents or toxins without authorisation/in violation of the conditions of an authorisation and penalties	6 (13%)	3 (8%)	7 (47%)	11 (35%)	4 (44%)	31 (23%)
Transfer dangerous biological agents or toxins without authorisation/to unauthorised persons and penalties	8 (18%)	12 (33%)	12 (80%)	14 (45%)	6 (67%)	52 (38%)

207

Appendix G

Measures	Africa: Number (%) of 45	Asia: Number (%) of 36	Eastern Europe: Number (%) of 15	Latin America and the Caribbean: Number (%) of 31	Western Europe and Other States: Number (%) of 9	Global: Number (%) of 136
CONTROL LISTS						
Control lists for dangerous biological agents and toxins	1 (2%)	7 (19%)	13 (87%)	3 (10%)	8 (89%)	32 (24%)
Control lists for dual-use biological equipment and related	1 (2%)	5 (14%)	11 (73%)	2 (6%)	4 (44%)	23 (17%)
PREVENTATIVE MEASURES						
Measures to account for production	0 (0%)	4 (11%)	1 (7%)	3 (10%)	2 (22%)	10 (7%)
Measures to account for use	0 (0%)	3 (8%)	4 (27%)	4 (13%)	4 (44%)	15 (11%)
Measures to account for storage	0 (0%)	3 (8%)	2 (13%)	1 (3%)	3 (33%)	9 (7%)
Measures to account for transport	1 (2%)	3 (8%)	3 (20%)	6 (19%)	4 (44%)	17 (13%)

Measures	Africa: Number (%) of 45	Asia: Number (%) of 36	Eastern Europe: Number (%) of 15	Latin America and the Caribbean: Number (%) of 31	Western Europe and Other States: Number (%) of 9	Global: Number (%) of 136
Measures to secure production	1 (2%)	3 (8%)	2 (13%)	0 (0%)	1 (11%)	7 (5%)
Measures to secure use	1 (2%)	3 (8%)	3 (20%)	2 (6%)	2 (22%)	11 (8%)
Measures to secure storage	1 (2%)	4 (11%)	4 (27%)	3 (10%)	1 (11%)	13 (10%)
Measures to secure transport	4 (9%)	4 (11%)	6 (40%)	13 (42%)	3 (33%)	30 (22%)
Regulations for physical protection of facilities which produce, use or store dangerous biological agents or toxins and related penalties	0 (0%)	3 (8%)	1 (7%)	1 (3%)	1 (11%)	6 (4%)
Regulations for physical protection of dangerous biological agents and toxins and related penalties	0 (0%)	3 (8%)	1 (7%)	0 (0%)	3 (33%)	7 (5%)
Authorisation of activities involving dangerous biological agents or toxins	5 (11%)	7 (19%)	6 (40%)	15 (48%)	6 (67%)	39 (29%)
National licensing authority	4 (9%)	4 (11%)	5 (33%)	12 (39%)	3 (33%)	28 (21%)
Regulations for genetic engineering work	10 (22%)	7 (19%)	10 (67%)	15 (48%)	6 (67%)	48 (35%)
TRANSFER CONTROLS						
Authorisation for exports and imports of dangerous biological agents and toxins	17 (38%)	18 (50%)	13 (87%)	24 (77%)	7 (78%)	79 (58%)
Export/import control authority	2 (4%)	12 (33%)	13 (87%)	16 (52%)	7 (78%)	50 (37%)
End-user controls for dangerous biological agents and toxins	2 (4%)	4 (11%)	13 (87%)	2 (6%)	6 (67%)	27 (20%)
Transit control over dangerous biological agents and toxins	3 (7%)	9 (25%)	13 (87%)	4 (13%)	4 (44%)	33 (24%)
Trans-shipment control over dangerous biological agents and toxins	0 (0%)	4 (11%)	0 (0%)	1 (3%)	2 (22%)	7 (5%)
Re-export control over dangerous biological agents and toxins	1 (2%)	6 (17%)	7 (47%)	2 (6%)	2 (22%)	18 (13%)
Export control over dangerous biological agents and toxins	9 (20%)	10 (28%)	13 (87%)	9 (29%)	6 (67%)	47 (35%)
Import control over dangerous biological agents and toxins	15 (33%)	12 (33%)	12 (80%)	14 (45%)	5 (56%)	58 (43%)

(Distributed by the Office of International Information Programs, U.S. Department of State. http://usinfo.state.gov.)

Notes

1. THE ROAD TO THE BIOLOGICAL WEAPONS CONVENTION (BWC)

1. A useful synopsis of arms control is provided at http://legal-dictionary.thefreedictionary.com/Arms+Control+and+Disarmament.
2. Barry Kolodkin, "What Is Arms Control?" http://usforeignpolicy.about.com/od/defense/a/what-is-arms-control.htm (accessed April 1, 2012).
3. Jonathan B. Tucker, *Innovation, Dual Use, and Security* (Cambridge: Massachusetts Institute of Technology, 2012), 49.
4. http://www.armscontrol.org/act/1997_04/cwctext (accessed April 1, 2012).
5. http://www.armscontrol.org/treaties/bwc (accessed April 1, 2012).
6. Jeffrey J. Byrd, PhD, and Tabitha M. Powledge, *The Complete Idiot's Guide to Microbiology* (New York: Penguin Group, 2006), 7.
7. Jenner's work with smallpox vaccination actually occurred on May 14, 1796, when Jenner tested his hypothesis by inoculating an eight-year-old boy named James Phipps with pus scraped from the cowpox blisters on the hands of Sarah Nelmes, a milkmaid who had caught cowpox.
8. The dissertation by Forrest Russel Frank, *U.S. Arms Control Policymaking: The 1972 Biological Weapons Convention Case*, Stanford University Press, 1975, provides a thorough description of the period leading up to the 1969 decision.
9. George Merck, "Official Report on Biological Warfare," *Bulletin of Atomic Scientists* 2, no. 9 (October 1946): 11–18.
10. "FAS Asks Clarification of U.S. Policies on Germ Warfare," *Bulletin of Atomic Scientists* 8, no. 6 (June 1952): 130.
11. Frank, *U.S. Arms Control Policymaking*, 34.
12. National Security Council, "Basic National Security Policy," NSC 5602/1, March 15, 1956, reprinted in U.S. Department of State, *Foreign Relations of the United States (FRUS), 1955 – 1957* 19, National Security Policy, 246.
13. Jeanne Guillemin, *Biological Weapons: From the Invention of State-Sponsored Programs to Contemporary Bioterrorism* (New York: Columbia University Press, 2005), 113.
14. *Public Papers of the Presidents of the United States: Dwight D. Eisenhower, 1960 – 1961* (Washington, DC: U.S. Government Printing Office, 1962), 62.
15. Frank, *U.S. Arms Control Policymaking*, 35.
16. Frank, *U.S. Arms Control Policymaking*, 55.

17. Joshua Lederberg, "A Treaty on Germ Warfare," *Washington Post*, September 24, 1966.

18. Frank, 64.

19. The source documents from this period have been complied and are available at http://www.gwu.edu/~nsarchiv/NSAEBB/NSAEBB58/ (accessed April 1, 2012).

20. Memorandum, Secretary of Defense Laird to National Security Advisor Henry Kissinger, April 30, 1969. Confidential, 1 p. Source: NARA, Nixon Presidential Materials, NSC Files, Subject Files, Box 310, Chemical, Biological Warfare (Toxins, etc.), Vol. I., http://www.gwu.edu/~nsarchiv/NSAEBB/NSAEBB58/RNCBW1.pdf.

21. National Security Study Memorandum (NSSM) 59, U.S. Policy on Chemical and Biological Warfare and Agents, from National Security Advisor Henry A. Kissinger to the Secretary of State, Secretary of Defense, Director of Central Intelligence, the Special Assistant to the President for Science and Technology, and the Director, U.S. Arms Control and Disarmament Agency, May 29, 1969, http://www.gwu.edu/~nsarchiv/NSAEBB/NSAEBB58/RNCBW4.pdf.

22. Frank, *U.S. Arms Control Policymaking*, 115–16.

23. Frank, *U.S. Arms Control Policymaking*, 115.

24. NSSM 59, November 10, 1969, 19.

25. Jonathan B. Tucker, and Erin R. Mahan, *President Nixon's Decision to Renounce the U.S. Offensive Biological Weapons Program*, Center for the Study of Weapons of Mass Destruction, National Defense University, Washington, DC, 5.

26. Report to the National Security Council, "US Policy on Chemical and Biological Warfare and Agents," submitted by the Interdepartmental Political-Military Group in response to NSSM 59, November 10, 1969. Top Secret, 52 pp., http://www.gwu.edu/~nsarchiv/NSAEBB/NSAEBB58/RNCBW6a.pdf.

27. NSSM 59, November 10, 1969, 20.

28. Ibid.

29. Lieutenant Colonel George W. Christopher, USAF, MC; Lieutenant Colonel Theodore J. Cieslak, MC, USA; Major Julie A. Pavlin, MC, USA; and Colonel Edward M. Eitzen Jr., MC, USA, "Biological Warfare: A Historical Perspective," Operational Medicine Division, United States Army Medical Research Institute of Infectious Diseases, Fort Detrick, Maryland.

30. "Memorandum, Presidential Science Advisor Lee A. DuBridge to National Security Advisor Henry Kissinger, October 22, 1969," Top Secret, 3 pp. NARA, Nixon Presidential Materials, NSC Files, Subject Files, Box 310, Chemical, Biological Warfare (Toxins, etc.), Vol. I, http://www.gwu.edu/~nsarchiv/NSAEBB/NSAEBB58/.

31. Ibid., 2.

32. Christopher et al., "Biological Warfare: A Historical Perspective," 6.

33. The cause-effect between the experiments and the Stanford University outbreak of urinary tract infections was never established/could have been unrelated—an early example of hospital acquired gram-negative bacterial infections complicating urinary tract instrumentation and antibiotic use (V. L. Yu, "*Serratia marcessens*: Historical Perspective and Clinical Review," *New England Journal of Medicine* 300 [1977; 1979]: 887–93). Serratia infections are common in modern hospitals in the absence of simulant studies-effects of surgery and antibiotic selection pressure.

34. BBC News, "Hidden History of U.S. Germ Testing," http://news.bbc.co.uk/2/hi/programmes/file_on_4/4701196.stm.

35. FAS.org, "Biological Weapons," http://www.fas.org/nuke/guide/usa/cbw/bw.htm.

36. Leonard A. Cole, *The Eleventh Plague: The Politics of Biological and Chemical Warfare* (New York: Macmillan, 2002), 19–23.

37. From undated paper by Peter J. Roman that references "Memorandum, George Kistiakowsky to James Killian, Subject: Biological Warfare," July 20, 1958.

38. Richard D. McCarthy, *The Ultimate Folly* (New York: Alfred A. Knopf, 1969), 62.

39. Frank, *U.S. Arms Control Policymaking*, 57 (from an interview with a former U.S. government official, November 1973).

40. NSSM 59, November 10, 1969, 18.

41. NSSM 59, November 10, 1969, 25.

42. Minutes of National Security Council Meeting, Washington, D.C., November 18, 1969, 2–3, in FRUS, document 161, available at www.state.gov/r/pa/ho/frus/nixon/e2/83592.htm. In attendance were President Nixon, Vice President Spiro Agnew, National Security Advisor Kissinger, Secretary of State William Rogers, Secretary of Defense Laird, Chairman of the Joint Chiefs of Staff General Wheeler, U.S. Arms Control and Disarmament Agency Director Gerard Smith, Director of Central Intelligence Helms, and Presidential Science Advisor Du-Bridge.

43. NSSM 59, November 10, 1969, 10.

44. NSSM 59, November 10, 1969, 8.

45. Ken Alibek, *Biohazard* (New York: Random House, 1999).

46. "Bio-Terrorism," Parliamentary Office of Science and Technology, United Kingdom, *Postnote* 166 (November 2001): 2, http://www.parliament.uk/post/pn166.pdf (May 16, 2008).

47. Alibek, *Biohazard*, 160.

48. The Nunn-Lugar program includes the development of programs for the former Soviet scientists that provide for using their knowledge for peaceful purposes and not returning to the development of BW.

49. S. Harris, "The Japanese Biological Warfare Programme: An Overview," in E. Geissler and J. E. van Courtland Moon (eds.), *Biological and Toxin Weapons: Research, Development and Use from the Middle Ages to 1945* (Oxford: Oxford University Press, 1999), 127–52.

50. The lack of a deterrent value of BW was also cited in NSSM 59, 25.

51. Tucker, *Innovation, Dual Use and Security*, 10.

52. Ibid., 102–3.

53. Jessica Stern, "Dreaded Risks and Control of Biological Weapons," *International Security* 27, 3 (Winter 2002–2003): 102–7.

54. The seventeen countries included Bulgaria, China, Cuba, Egypt, India, Iran, Iraq, Israel, Laos, Libya, North Korea, Russia, South Africa, South Korea, Syria, Taiwan, and Vietnam. (Cordesman, Anthony H., The Challenge of Biological Terrorism, CSIS Press: Washington, DC, 2005, p. 15.)

55. Minutes of National Security Council Meeting, Washington, D.C., November 18, 1969, 4.

56. President Richard Nixon, "Statement on Chemical and Biological Defense Policies and Programs, November 25, 1969," Public Papers of the Presidents (Washington, DC: U.S. Government Printing Office, 2004), 968–69.

57. Tucker and Mahan, *President Nixon's Decision to Renounce the U.S. Offensive Biological Weapons Program*, 16.

58. Ibid.

2. BIOTECHNOLOGY AND ITS IMPLICATIONS FOR THE BWC

1. The Industrialization of Biology and Its Impact on National Security, University of Pittsburg Medical Center (UPMC), June 8, 2012, http://www.upmc-biosecurity.org/website/resources/publications/2012/pdf/2012-06-08-industrialization_bio_natl_security.pdf.

2. Joseph R. Ecker, Wendy A. Bickmore, Inês Barroso, Jonathan K. Pritchard, Yoav Gilad, and Eran Segal. "Genomics: ENCODE Explained," *Nature* 489, nos. 52–55 (September 6, 2012), doi:10.1038/489052a, http://www.nature.com/nature/journal/v489/n7414/full/489052a.html.

3. Gina Kolata, "Bits of Mystery DNA, Far from 'Junk,' Play Crucial Role," *New York Times*, Science Section, September 5, 2012, http://www.nytimes.com/2012/09/06/science/far-from-junk-dna-dark-matter-proves-crucial-to-health.html?emc=eta1.

4. Ecker et al., "Genomics: ENCORE Explained," ibid.

5. Jonathan B. Tucker, Innovation, Dual Use, and Security (Cambridge: Massachusetts Institute of Technology, 2012), 4.

6. Tucker, 5.

7. Kevin Keener, Thomas Hoban, and Rekha Balasubramanian, "Biotechnology and Its Applications," North Carolina State University, Department of Food Science, 4, http://www.ces.ncsu.edu/depts/foodsci/ext/pubs/bioapp.html.

8. *The Industrialization of Biology and Its Impact on National Security*, 4.

9. Tucker, 3.

10. Keener, Hoban, and Balasubramanian, "Biotechnology and Its Applications," 2.

11. Tucker, 200.

12. *The Industrialization of Biology and Its Impact on National Security*, 15.

13. Robert Carlson, "Biodefense Net Assessment: Causes and Consequences of Bioeconomic Proliferation: Implications for U.S. Physical and Economic Security." Homeland Security Studies and Analysis Institute publication prepared for the Department of Homeland Security (BNA 2012-03), http://www.google.com/search?sourceid=navclient&ie=UTF-8& rlz=1T4RNSN_enUS400US401& q=Biodefense+Net+Assessment%3a+Causes+and+Consequences+of+Bioeconomic+Proliferat ion%3a+Implications+for+U.S.+Physical+and+Economic+Security.++Homeland+Security+St udies+and+Analysis+Institute+publication+prepared+for+the+Department+of+Homeland+Sec urity+%28BNA+2012-03%29 (accessed October 20, 2012).

14. Tucker, 24.

15. Tucker, 25.

16. The concept of BioBricks "envisions the design and construction of genetic circuits and modules using a tool kit of DNA sequences" (Tucker, 25).

17. *The Industrialization of Biology and Its Impact on National Security*, 7.

18. Freeman Dyson, "Our Biotech Future," *New York Review of Books* 54, no. 12 (July 19, 2007), http://www.nybooks.com/articles/archives/2007/jul/19/our-biotech-future.

19. M. Curriu, J. Carrillo, N. Massanella, E. Garcia, F. Cunyat et al., "Susceptibility of Human Lymphoid Tissue Cultured ex vivo to Xenotropic Murine Leukemia Virus-Related Virus (XMRV) Infection." *PLoS ONE* 7(5): e37415, doi:10.1371/journal.pone.0037415.

20. Tucker, 135.

21. United States Department of Agriculture website at http://www.invasivespeciesinfo.gov/economic/main.shtml.

22. Tucker, 151.

23. Meredith Cohn, "Gene Patent Case Could Impact Patients, Research," Baltimore-Sun.com, http://www.baltimoresun.com/health/bs-hs-brca-patents-20120922,0,2138615,full. story.

24. The definition of synthetic biology comes from Tucker, 20.

25. http://en.wikipedia.org/wiki/Polymerase_chain_reaction (accessed October 20, 2012).

26. *The Militarily Critical Technologies List Part II: Weapons of Mass Destruction Technologies (ADA 330102)*, "Biological Weapons Technology," U.S. Department of Defense, Office of the Under Secretary of Defense for Acquisition, Logistics and Technology, February 1998. http://www.fas.org/irp/threat/mctl98-2/p2sec03.pdf.

3. TEN REASONS WHY THE BWC MATTERS

1. Presidential Policy Directive PPD/8 National Preparedness Overview Briefing Effective January 11, 2012, http://www.emforum.org/vforum/FEMA/PPD-8.pdf.

2. Parliamentary Office of Science and Technology, "Bio-Terrorism," *Postnote* 166 (November 2001): 2, http://www.parliament.uk/post/pn166.pdf (accessed May 16, 2008).

3. Ibid.

4. From a briefing delivered to the National Security Staff titled "US Biological Weapons Program: Implications for US Biodefense in 2010." Delivered by Mr. Joel McCleary, December 2, 2010.

5. Michael Christopher Carroll, *Lab 257 : The Disturbing Story of the Government's Secret Plum Island Germ Laboratory* (New York: Harper Collins, 2005), 14–15.

6. See Graham S. Pearson, "Biological Weapons: Their Nature and Arms Control," in *Non-Conventional Weapons Proliferation in the Middle East: Tackling the Spread of Nuclear, Chemical, and Biological Capabilities*, edited by Efraim Karsh, Martin S. Navias, and Philip Sabin (Oxford: Clarendon Press, 1993), 113.

7. "Secret Project Manufactured Mock Anthrax," *Washington Times*, October 26, 2001, http://www.washingtontimes.com/news/2001/oct/26/20011026-030448-2429r/.

8. National Strategy for Countering Biological Threats (http://www.hsdl.org/?view& did=31404) and the Presidential Policy Directive-2 for implementing the Strategy (http:// www.fas.org/irp/offdocs/ppd/ppd-2.pdf) (accessed April 29, 2012).

9. From a National Security Staff briefing "Countering Biological Threats" (undated) concerning the development of the PPD-2.

10. Joseph Cirincione, Jon B. Wolfsthal, and Miriam Rajkumar, *Deadly Arsenals: Nuclear, Biological, and Chemical Threats* (Washington, DC: Carnegie Endowment for International Peace, 2005), 61.

11. Ken Alibek, *Biohazard* (New York: Random House, 1999), 259–61.

12. Guillemin, Jeanne. *Biological Weapons: From the Invention of State-Sponsored Programs to Contemporary Bioterrorism*. New York: Columbia University Press, 2005, pp. 107.

13. David Hoffman, *The Dead Hand: The Untold Story of the Cold War Arms Race and Its Dangerous Legacy* (New York: Anchor Books, 2010).

14. Joshua Lederberg, "Introduction," in *Biological Weapons: Limiting the Threat*, edited by Joshua Lederberg (Cambridge, MA: MIT Press, 1999), 6.

15. Jonathan B. Tucker, Innovation, Dual Use, and Security (Cambridge: Massachusetts Institute of Technology, 2012), 32-35.

16. Phillip M. McCauley and Rodger A. Payne, "The Illogic of the Biological Weapons Taboo," *Strategic Studies Quarterly* 4 (Spring 2010): 21–22, 27–28.

17. In Raymond M. Zilinskas, ed. *Biological Warfare: Modern Offense and Defense* (Boulder, CO: Lynne Reiner, 2000), 150, the author cites 142 parties and another eighteen nations that have signed the treaty but not ratified it.

18. Anthony H. Cordesman, *The Challenge of Biological Terrorism* (CSIS Press: Washington, DC, 2005), 15.

19. Cirincione et al., *Deadly Arsenals*, 11.

20. Zilinskas, ed., *Biological Warfare*, 150.

21. Cordesman, *The Challenge of Biological Terrorism*, 15. The seventeen countries included Bulgaria, China, Cuba, Egypt, India, Iran, Iraq, Israel, Laos, Libya, North Korea, Russia, South Africa, Syria, Taiwan, and Vietnam.

22. Jessica Stern, "Dreaded Risks and Control of Biological Weapons," *International Security* 3 (2003): 102–7.

23. *Globalization, Biosecurity, and the Future of the Life Sciences*. Institute of Medicine and the National Research Council of the National Academies (Washington, DC: National Academies Press, 2006), 50–51.

24. Tim Lister, and Paul Cruickshank, "From the Grave, al-Awlaki Calls for Bio-chem Attacks on the U.S." CNN Report, May 2, 2012, http://security.blogs.cnn.com/2012/05/02/ from-the-grave-al-awlaki-calls-for-bio-chem-attacks-on-the-u-s/.

25. http://www.australiagroup.net/en/objectives.html

26. http://www.state.gov/t/isn/c10390.htm

27. Ibid.

28. http://www.who.int/about/en/

29. http://www.fao.org/partnerships/partner-un/en/

30. http://www.oie.int/about-us/our-missions/

31. Ibid.

32. "The FAO-OIE-WHO Collaboration: Sharing Responsibilities and Coordinating Global Activities to Address Health Risks at the Animal-Human-Ecosystems Interfaces—A Tripartite Concept Note," April 2010, http://www.oie.int/fileadmin/Home/eng/Current_Scientific_Issues/ docs/pdf/FINAL_CONCEPT_NOTE_Hanoi.pdf.

33. Computer chips have been doubling in processing speed every eighteen months for several decades. This phenomenon is known as Moore's Law after semiconductor pioneer

Gordon Moore, who first predicted it in 1965. This constant increase in processing speed has helped fuel the microelectronic and information revolutions.

34. Randall Mayes, "Book Review: Robert Carlson's *Biology Is Technology: The Promise, Peril, and New Business of Engineering Life,*" *Journal of Evolution and Technology* 21, no. 1 (June 2010): 55–59, http://jetpress.org/v21/mayes.htm.

35. http://crossborderbiotech.ca/trends-in-2009/

36. William W. Thompson, Lorraine Comanor, and David K. Shay, "Epidemiology of Seasonal Influenza: Use of Surveillance Data and Statistical Models to Estimate the Burden of Disease," *Journal of Infectious Disease* 194, Supplement 2 (2006): S82–S91. doi: 10.1086/507558, http://jid.oxfordjournals.org/content/194/Supplement_2/S82.full.pdf+html.

37. Ron Plain, University of Missouri, Economist, May 24, 2010.

38. IOM Microbial Threats 2006. (www.nap.edu/openbook.php?record_id=12799&page=297).

39. From a National Security Staff briefing "Countering Biological Threats" (undated) concerning the development of the PPD-2.

40. The Non-Aligned Movement (NAM) is a group of states that are not aligned formally with or against any major power bloc. As of 2012, the movement has 120 members and seventeen observer countries.

41. National Strategy for Countering Biological Threats, National Security Council, The White House, November 2009.

42. From University of Pittsburg Medical Center website, http://www.upmc-biosecurity.org/website/events/2001_darkwinter/findings.html (accessed May 4, 2012).

43. Bruggen, Koos van der and Barend ter Haar. The Future of Biological Weapons Revisited. Netherland Institute for International Relations, December 2011, p. 39.

44. United Nations Security Council Functions, http://www.un.org/docs/sc/unsc_functions.html (accessed May 12, 2011).

45. http://www.un.org/en/sc/1540/ (accessed May 12, 2012).

46. United Nations Security Council Resolution 1441 is a United Nations Security Council resolution adopted unanimously by the United Nations Security Council on November 8, 2002, offering Iraq under Saddam Hussein "a final opportunity to comply with its disarmament obligations" that had been set out in several previous resolutions. http://en.wikipedia.org/wiki/United_Nations_Security_Council_Resolution_1441 (accessed May 12, 2012).

47. D. P. Fidler, and L. O. Gostin, "The Securitization of Public Health" in D. P. Fidler and L. O. Gostins (eds.), *Biosecurity in the Global Age: Biological Weapons, Public Health and the Rule of Law* (Stanford: Stanford University Press, 2008), 136.

48. The information on BW as an environmental issues comes from *BioWeapons Prevention Project: Biological Weapons Reader*, edited by Kathryn McLaughlin and Kathryn Nixdorff (Geneva, 2009), chapter 7, "Biological Weapons as a Environmental Issue," 61.

49. Ibid., 63–65.

50. Ibid., 68.

51. *Biotechnology Research in an Age of Terrorism* (Washington, DC: The National Academies Press, 2001), http://www.nap.edu/catalog/10827.html.

52. P. A. Gilbert, and G. McFadden, "Poxvirus Cancer Therapy," *Recent Patents on Anti-Infective Drug Discovery* 1 (2006): 309–21.

53. NSABB, "Proposed Framework for the Oversight of Dual Use Life Sciences Research: Strategies for Minimizing the Potential Misuse of Research Information," National Institutes of Health, June 2007, http://oba.od.nih.gov/biosecurity/pdf/Framework%20for%20transmittal%200807_Sept07.pdf.

54. From the London-based Verification, Research, Training and Information Centre (VERTIC) as provided to the author. The VERTIC website is http://www.vertic.org/pages/homepage/programmes/national-implementation-measures/biological-weapons-and-materials/bwc-legislation-database/introduction.php (accessed April 8, 2012).

4. INTO THE ABYSS: THE ARTICLES OF THE CONVENTION

1. NSSM 59, November 10, 1969, 18.

2. From a BWC workshop titled, "Trends in Science and Technology Relevant to the Biological and Toxin Weapons Convention," held October 31–November 3, 2010, at the Institute of Biophysics of the Chinese Academy of Sciences and convened under the auspices of IAP—The Global Network of Science Academies, the International Union of Biochemistry and Molecular Biology (IUBMB), the International Union of Microbiological Societies (IUMS), the Chinese Academy of Sciences, and the U.S. National Academies.

3. Koos van der Bruggen, and Barend ter Haar, *The Future of Biological Weapons Revisited* (The Hague: Netherland Institute for International Relations, 2011), 45.

4. Christian Enemark, *Disease and Security: Natural Plagues and Biological Weapons in East Asia* (New York: Routledge, 2007), 173–75; and Judith Miller, Stephen Engelberg, and William J. Broad, "U.S. Germ Warfare Research Pushes Treaty Limits," *New York Times*, September 4, 2001.

5. Jonathan B. Tucker, "Biological Threat Assessment: Is the Cure Worse Than the Disease?" *Arms Control Today*, October 2004.

6. Final 2011 BWC Review Conference document at http://daccess-dds-ny.un.org/doc/UNDOC/GEN/G12/600/60/PDF/G1260060.pdf?OpenElement, 10.

7. www.drcordas.com/education/weaponsmassd/Saxitoxin.ppt (accessed April 8, 2012).

8. Final 2011 BWC Review Conference document, 10.

9. From the BWC Confidence Building Measures adopted in 1986.

10. From the London-based Verification, Research, Training and Information Centre (VERTIC) as provided to the author. The VERTIC website is http://www.vertic.org/pages/homepage/programmes/national-implementation-measures/biological-weapons-and-materials/bwc-legislation-database/introduction.php (accessed April 8, 2012).

11. In Raymond M. Zilinskas, ed., *Biological Warfare: Modern Offense and Defense* (Boulder, CO: Lynne Reiner, 2000), 150, the author cites 142 parties and another eighteen nations that have signed the treaty but not ratified it.

12. Anthony H. Cordesman, *The Challenge of Biological Terrorism* (CSIS Press: Washington, DC, 2005), 15.

13. Cordesman, *The Challenge of Biological Terrorism*, 15. The seventeen countries included Bulgaria, China, Cuba, Egypt, India, Iran, Iraq, Israel, Laos, Libya, North Korea, Russia, South Africa, South Korea, Syria, Taiwan, and Vietnam.

14. Cirincione, Joseph, et al., Deadly Arsenals, Tracking Weapons of Mass Destruction, rev. ed. (Washington, DC: Carnegie Endowment for International Peace, 2005), pp. 11.

15. Zilinskas, *Biological Warfare*, 150.

16. Final 2011 BWC Review Conference document, 13.

17. http://www.armscontrol.org/treaties/bwc (accessed April 1, 2012).

18. Ibid., 14.

19. Ibid.

20. *The Militarily Critical Technologies List Part II: Weapons of Mass Destruction Technologies (ADA 330102)*, "Biological Weapons Technology," U.S. Department of Defense, Office of the Under Secretary of Defense for Acquisition, Logistics and Technology, February 1998, http://www.fas.org/irp/threat/mctl98-2/p2sec03.pdf.

21. Carlson, Rob Carlson, "The Proliferation of Biotechnologies," *Biosecurity and Bioterrorism: Biodefense Strategy, Practice and Science* 1, no. 3 (August 2003): 1–3.

22. Milton Leitenberg, *Assessing the Biological Weapons and Bioterrorism Threat* (Carlisle, PA: Strategic Studies Institute, U.S. Army War College, 2005), 35.

23. Testimony Before the Committee on Homeland Security and Governmental Affairs United States Senate, "Dual Use Research of Concern: Balancing Benefits and Risks," Statement of Anthony S. Fauci, M.D., Director National Institute of Allergy and Infectious Diseases National Institutes of Health U.S. Department of Health and Human Services, http://www.niaid.nih.gov/about/directors/Documents/DURCfinalClearedTestimony.PDF, 1.

24. Koos van der Bruggen, and Barend ter Haar, "The Future of Biological Weapons Revisited: A Concise History of the Biological and Toxin Weapons Convention" (The Hague: Netherlands Institute of International Relations, 2011), 15.

25. Data is current as of April 2012. Numbers come from the BWC and CWC websites, respectively, at http://www.opbw.org/ and http://www.opcw.org/chemical-weapons-convention/ (accessed April 27, 2012).

26. A listing of Declarations and Reservations can be found on the BWC website at http://www.opbw.org/convention/documents/btwcres.pdf (accessed April 27, 2012).

27. Van der Bruggen and ter Haar, "The Future of Biological Weapons Revisited," 31.

5. THE BWC REVIEW CONFERENCES (REVCON) AND SPECIAL REVIEWS

1. "The Seventh BWC Review Conference: Setting the Scene," RevCon Report #1, Monday, December 5, 2011, 1, http://www.bwc2011.info/other-docs/RC11-01.pdf (accessed April 27, 2012).

2. See the United Nations in Geneva website at http://www.unog.ch/80256EE600585943/ (httpPages)/92CFF2CB73D4806DC12572BC00319612?OpenDocument.

3. http://en.wikipedia.org/wiki/Asilomar_conference_on_recombinant_DNA

4. Review Conference to the Parties to the Convention on the Prohibition on the Development, Production and Stockpiling of Bacteriological (Biological) and Toxin Weapons and on Their Destruction: Final Document, Geneva, 1980, 12, http://bwc.unog.ch/1980-03-1RC/BWC_CONF.I_10.pdf.

5. Ibid., 9, http://bwc.unog.ch/1980-03-1RC/BWC_CONF.I_10.pdf.

6. Attacks were never witnessed by Western intelligence operatives, and direct samples of aerosols were not recovered. Residues collected from environmental surfaces were similar to naturally occurring showers of bee feces in microscopic appearance, pollen, and fungal content. Mycotoxin was found in less than 10 percent of the samples, with widely varying results coming from different laboratories. Problems confounding the analysis included contradictory accounts from alleged victims, low attack rates, and highly variable symptoms in allegedly exposed individuals (T. D. Seeley et al., "Yellow Rain," *Scientific American* 253, no. 3 [1985]: 128–37.

7. Koos van der Bruggen, and Barend ter Haar, *The Future of Biological Weapons Revisited: A Concise History of the Biological and Toxin Weapons Convention* (The Hague: Netherlands Institute for International Relations, December 2011), 38.

8. Ibid., 40.

9. http://bwc.unog.ch/1986-09-2RC/BWC_CONF.II_02.pdf, 3

10. Van der Bruggen and ter Haar, *The Future of Biological Weapons Revisted*, 43.

11. Briefing Book, Seventh BWC Review Conference, http://www.bwc2011.info/, 47.

12. Ibid., 49.

13. Ibid., 64.

14. Ibid.

15. Ibid., 72.

16. Ibid., 249.

17. Ad Hoc Group of Governmental Experts to Identify and Examine Potential Verification Measures from a Scientific and Technical Standpoint. Second Session, Geneva (November 23–December 4, 1992), 2, http://www.opbw.org/verex/docs/sess2/conf_docs/BWC_CONF.III_VEREX_4.pdf.

18. Ad Hoc Group of Governmental Experts to Identify and Examine Potential Verification Measures from a Scientific and Technical Standpoint. Second Session, Geneva (September 13–24, 1993), Briefing Book, Seventh BWC Review Conference, http://www.bwc2011.info/, 251.

19. Ibid., 253.

20. Van der Bruggen and ter Haar, *The Future of Biological Weapons Revisted*, 88.

21. Ibid., 93.

22. Ad Hoc Group of Governmental Experts to Identify and Examine Potential Verification Measures from a Scientific and Technical Standpoint, Twenty-Second Session, Geneva, February 12–23, 2001, BWC/AD HOC GROUP/55-1, 4, http://www.opbw.org/ahg/docs/rolling%20text%20and%20annexes.pdf.

23. Van der Bruggen and ter Haar, *The Future of Biological Weapons Revisted*, 95.

24. John R. Bolton, Under Secretary of State for Arms Control, Tuesday, November 20, 2001. Remarks to the Fifth Review Conference, http://archive.newsmax.com/archives/articles/2001/11/19/170241.shtml.

25. Briefing Book, Seventh BWC Review Conference, http://www.bwc2011.info/, 95–96.

26. Ibid., 109.

27. Ibid., 118.

28. Ibid., 120.

29. UNOG website, http://www.unog.ch/80256EE600585943/(httpPages)/F1CD974A1FDE4794C125731A0037D96D?OpenDocument.

30. Tim Farnsworth, "Don't Neglect the Biological Weapons Convention," *Arms Control Now*, January 12, 2012.

31. Ibid., 3.

32. Bioweapons Prevention Project RevCon Report #16: The Seventh BWC Review Conference: Outcome and Assessment, Geneva, December 31, 2011, 2, http://www.bwpp.org/documents/RC11-16.pdf.

33. Ibid.

34. Ibid.

6. TO VERIFY OR NOT TO VERIFY

1. Amy F. Woolf, *Monitoring and Verification in Arms Control*, Congressional Research Service, Washington, D.C., December 23, 2011, 3.

2. Ibid., 5–6.

3. Congressional Report, 111th Congress, Threat with Russia on Measures for Further Reduction and Limitation of Strategic Offensive Arms (The New START Treaty), October 1, 2010, 21.

4. John G. Tower, James Brown, and William K. Cheek, *Verification: The Key to Arms Control in the 1990s* (Washington, DC: Brassey's, 1992), xviii.

5. Amy F. Woolf, *Verification: The Critical Element of Arms Control*, U.S. Arms Control and Disarmament Agency, Annual Report to Congress, 1988, Washington, DC, 1989, 55.

6. Raymond A. Zilinskas, "Take Russia to 'Task' on Bioweapons Transparency," *Nature Medicine* 18, no. 850 (June 2012).

7. Accounts of the tripartite inspections and agreement were contained in David C. Kelly, "The Trilateral Agreement: Lessons for Biological Weapons Verification," *Verification Yearbook 2002*, 93–110.

8. Ken Alibek, *Biohazard* (New York: Random House, 1999), 225–40.

9. Joint U.S./UK/Russian Statement on Biological Weapons, September 14, 1992, at http://www.fas.org/nuke/control/bwc/text/joint.htm (accessed June 20, 2012).

10. Trevor Findlay, and Oliver Meier (eds.), *VERTIC Verification Yearbook 2002*, , section by David Kelly, "The Trilateral Agreement: Lessons for Biological Weapons Verification," 100, http://www.isn.ethz.ch/isn/Digital-Library/Publications/Detail/?ots591=0c54e3b3-1e9c-be1e-2c24-a6a8c7060233&lng=en&id=13529 (accessed December 8, 2012).

11. Ibid., 103–5.

12. Adherence to and Compliance with Arms Control, Nonproliferation, and Disarmament Agreements and Commitments, U.S. Department of State, August 2011, 13.

13. Amy E. Smithson, *Germ Gambits: The Bioweapons Dilemma, Iraq and Beyond* (Stanford: Stanford University Press, 2011), xi.

14. Ibid.
15. Ibid., 27.
16. Ibid., 44.
17. Ibid., 61.
18. Ibid., 73.
19. Assessment is provided at http://dtirp.dtra.mil/TIC/treatyinfo/vd_comp05.htm#chapter6 Accessed on March 24, 2013.

7. THE FUTURE OF THE BWC

1. This number does not include the hundreds of white-powder mails that occur annually in the United States alone or the use of compounds such as ricin in assassination attempts.
2. Tim Farnsworth, "Don't Neglect the Biological Weapons Convention," *Arms Control Now*, January 12, 2012.
3. Testimony Before the Subcommittee on Oversight and Investigations, Committee on Energy and Commerce, House of Representatives, High Containment Biosafety Laboratories: Preliminary Observations on the Oversight of the Proliferation of BSL-3 and BSL-4 Laboratories in the United States, Statement of Keith Rhodes, Chief Technologist, Center for Technology and Engineering Applied Research and Methods, GAO-08-108T, October 4, 2007, 9–10.
4. From http://daccess-dds-ny.un.org/doc/UNDOC/GEN/G12/600/60/PDF/G1260060.pdf? OpenElement, 6. States included: Afghanistan, Albania, Algeria, Argentina, Armenia, Australia, Austria, Azerbaijan, Bangladesh, Belarus, Belgium, Bhutan, Bosnia and Herzegovina, Botswana, Brazil, Brunei, Darussalam, Bulgaria, Burundi, Canada, Chile, China, Colombia, Costa Rica, Croatia, Cuba, Cyprus, Czech Republic, Democratic Republic of Congo, Denmark, Dominican Republic, Estonia, Ethiopia, Fiji, Finland, France, Georgia, Germany, Ghana, Greece, Guatemala, Holy See, Hungary, India, Indonesia, Iran (Islamic Republic of), Iraq, Ireland, Italy, Japan, Jordan, Kazakhstan, Kenya, Kuwait, Lao People's Democratic Republic, Latvia, Lebanon, Lesotho, Libya, Liechtenstein, Lithuania, Madagascar, Malaysia, Mexico, Mongolia, Morocco, Mozambique, Netherlands, New Zealand, Nigeria, Norway, Pakistan, Peru, Philippines, Poland, Portugal, Qatar, Republic of Korea, Republic of Moldova, Romania, Russian Federation, Saudi Arabia, Senegal, Serbia, Singapore, Slovakia, Slovenia, South Africa, Spain, Sri Lanka, Sweden, Switzerland, Tajikistan, Thailand, , Tunisia, Turkey, Uganda, Ukraine, United Arab Emirates, United Kingdom of Great Britain and Northern Ireland, United States of America, Uruguay, Venezuela (Bolivarian Republic of), Yemen, and the former Yugoslav Republic of Macedonia.
5. Amy F. Woolf, *Monitoring and Verification in Arms Control*, Congressional Research Service, Washington, D.C., December 23, 2011, 3.
6. National Strategy for Countering Biological Threats, National Security Council, November 2009, http://www.whitehouse.gov/sites/default/files/National_Strategy_for_Countering_BioThreats.pdf
7. James R. Ferguson, "Biological Weapons and US Law," *Journal of American Medicine* 278, no. 5 (August 6, 1997): 357–60, doi:10.1001/jama.1997.03550050017006.
8. Media Note, U.S. Mission in Geneva, BWC Review Conference—U.S. Highlights International Assistance, Need for Coordinated Preparedness and Response, December 8, 2011, from the briefing presented by Deputy Under Secretary Daniel M. Gerstein, Science and Technology Directorate, U.S. Department of Homeland Security, http://geneva.usmission.gov/wp-content/uploads/2011/12/Dec.8GersteinPresentation.pdf.
9. From IAEA website at http://www.iaea.org/About/index.html.
10. Available at http://www.atlantic-storm.org/flash/flash.htm (accessed June 8, 2012).
11. Daniel Feakes, "Evaluating the CWC Verification System," http://www.unidir.org/pdf/articles/pdf-art1822.pdf.
12. Daniel Horner, "CWC Members Debate Inspection Distribution," *Arms Control Today*, January/February 2011, http://www.armscontrol.org/act/2011_01-02/CWC.

13. Schedules categories from Wikipedia at http://en.wikipedia.org/wiki/Chemical_Weapons_Convention are as follows:

- Schedule 1 chemicals have few, or no, uses outside of chemical weapons. These may be produced or used for research, medical, pharmaceutical, or chemical weapon defense-testing purposes but production above one hundred grams per year must be declared to the OPCW. A country is limited to possessing a maximum of one tonne of these materials. Examples are mustard and nerve agents and substances that are solely used as precursor chemicals in their manufacture. A few of these chemicals have very small-scale nonmilitary applications; for example, minute quantities of nitrogen mustard are used to treat certain cancers.
- Schedule 2 chemicals have legitimate small-scale applications. Manufacture must be declared, and there are restrictions on export to countries that are not CWC signatories. An example is thiodiglycol, which can be used in the manufacture of mustard agents but is also used as a solvent in inks.
- Schedule 3 chemicals have large-scale uses apart from chemical weapons. Plants that manufacture more than thirty tonnes per year must be declared and can be inspected, and there are restrictions on the export to countries that are not CWC signatories. Examples of these substances are phosgene, which has been used as a chemical weapon but which is also a precursor in the manufacture of many legitimate organic compounds, and triethanolamine, used in the manufacture of nitrogen mustard but also commonly used in toiletries and detergents.

14. United Nations Monitoring, Verification and Inspection Commission (UNMOVIC) was created through the adoption of United Nations Security Council resolution 1284 of December 17, 1999.

15. "Winter Solstice—A Long Shortest Day: Working into the Night," BWPP RevCon Report #14, December 22, 2011, 2.

16. Ibid.

APPENDIX D: ARMS CONTROL DEFINITIONS

1. A useful synopsis of arms control is provided at http://legal-dictionary.thefreedictionary.com/Arms+Control+and+Disarmament.

2. Michael Krepon, Defining Arms Control, Armscontrolwonk.com, October 8, 2009, at http://krepon.armscontrolwonk.com/archive/2497/defining-arms-control (accessed June 22, 2012).

3. The National Security Strategy of the United States, The White House, 1990, Washington, D.C., 15–16, http://bushlibrary.tamu.edu/research/pdfs/national_security_strategy_90.pdf.

APPENDIX G: NATIONAL IMPLEMENTATION

1. Verification Research, Training and Information Centre, Development House, 56–64 Leonard Street, London, United Kingdom, website www.vertic.org.

Index

1918 Great Influenza, 9, 38, 39, 49, 66
3-dimensional (or 3-D) printing, 33, 43

Ad Hoc Group (AHG), 108, 111, 112, 113, 114–115, 116, 117, 195, 196, 199, 200, 201, 203, 204
advanced manufacturing, 33, 41
Afghanistan, 54, 107
aflatoxin, 141
Africa, 4, 34, 58, 59, 63, 65, 69, 94, 146, 164
African swine fever, 65
Age of Biotechnology, xi, xii, xiii, xiv, xvi, 29, 35, 45, 52, 64, 69, 76, 79, 81, 84, 120, 145, 154, 163, 171, 173, 177
agriculture, 1, 27, 34, 54, 61, 64, 65, 67, 79, 96, 138, 146, 151, 156
al-Awlaki, Anwar, 60
Alibekov, Kanatjan (also Ken Alibek), 21, 22, 56, 137
All-Union Scientific Research Institute of Veterinary Virology at Pokrov, 138
al-Qaeda, xii, 60, 115
American Civil Liberties Union, 45
American Indians (see also Native Americans), 78
Amerithrax, xii, 56, 69, 72, 73, 115, 116, 145, 148, 156
amino acids, 42
anthrax, xii, 7, 17, 22, 23, 24, 40, 47, 54, 55, 56, 57, 59, 69, 72, 73, 77, 88, 106,
107, 115, 136, 140, 141
antibiotics, 18, 22, 25, 39, 41, 56, 66, 80, 83, 107, 127
Aral Sea, 22
Arkansas, 16, 27
Arms Control and Disarmament Agency (ACDA), 12, 26
arms control, xi, xiii–xiv, xv, xv, xvi, xvii, 1–7, 13, 26, 27, 33, 43, 53, 54, 57, 60, 74, 75–76, 79, 81, 82, 84, 85, 86, 93, 95, 98, 101, 103, 106, 115, 116, 119, 120, 124, 128, 129, 130, 131, 132, 134, 134, 135–136, 139, 142, 144, 146, 147, 151–152, 153, 155, 173, 175–176, 177, 185; chemical (see also Chemical Weapons Convention (CWC), 2, 3, 4–5, 10–11, 11–12, 12–15, 26, 27, 33, 57, 76, 82, 85, 108, 110, 111, 118, 152, 163, 167, 168, 176; conventional, 2, 4, 19, 128, 133; definitions, xvi, 86, 128, 150, 151, 159, 160, 189–191; elements, xiii, xv, xvi, 2, 43, 74, 86, 101; history, 1–4, 43, 75, 76; measures, 57, 112–113, 189; nuclear, 1, 2, 3–4, 9, 11, 12, 24, 27, 74, 75, 76, 81, 82, 85, 95, 102, 118, 130, 132, 133, 147, 151, 154, 159, 176, 177; verification, ix, xiv, xv, xvi–xvii, 2, 3, 193–194
Asilomar Conference, 105
Aspergillus fumigatus, 17
Atlantic Storm exercise, 73, 165

223

Aum Shinrikyo, xii, 59
Australia Group, 61, 69, 75, 164, 197
Azerbaijan, 90, 164
Aziz, Tariq, 141

Bacillus anthracis , 17, 40, 77, 88, 90
Bacillus subtilis var. *globigii* , 17, 55, 77
Bacillus thuringiensis , 22
bacteria, 7, 16, 19, 22, 23, 30, 35, 36, 38, 40, 41, 47, 48, 56, 62, 71, 88, 141, 152, 160
ballistic missiles, 3, 21
Beijing, 120
Belarus, 164
Berlin, 120, 190
Berlin Wall, 190
Bilateral Consultative Commission (BCC), for "New START", 132
bioactive peptides, 41
BioBricks, 36, 37, 41
biodiversity, xiv, xvi, 62, 75, 78, 151, 174, 185
bioenergy, 62
bioethics, 44, 46
bioforensics, 73, 83
biofuels, 41, 152
biohacker, 43, 48
Biohazard, 137
bioinformatics, 31, 32, 33, 39, 41, 97, 152
biological weapons: acquisition of pathogen, 25, 38, 70, 74, 77, 82, 85, 87, 88, 89, 91, 103, 147, 156, 158, 159; as deterrence, 13, 20, 24, 26, 55, 134, 149, 199; context, 8–9; definition, 7; dreaded risks of, 9, 24, 58; incapacitating agent, 7, 10, 13, 14, 18, 20; international BW programs, 23–25; Iraq BW program, 59, 90, 115, 140; Japanese BW program, 15, 23, 59; lethal agent, 7, 11, 13, 14, 19, 22, 37, 40, 55, 66, 68; norm against BW, xiv, xv, 57–60, 84, 85, 86, 103, 136, 149, 150, 161, 171, 175, 185, 199, 205; processing of pathogen, 8, 22, 25, 35, 36, 40, 41, 43, 55, 70, 79, 81, 132, 152; Soviet offensive program, xii, 20–23, 47, 56–57, 58, 59, 69, 73, 74, 88, 89, 90, 94, 106, 107–108, 136–139, 149, 159, 164; strategic weapons, xi, xii, 11, 16, 18, 19, 20, 21, 23, 46, 47,

56, 57, 81, 84; U.S. concepts for employment, 9–15; U.S. offensive program, 16–20; weaponization, 16–18, 55, 81, 89, 127, 141
Biopreparat, 21, 56, 136, 137
bioregulators, 41, 42, 128
biorisk, 44, 46
biosafety, xiv, 44, 46, 93, 118, 152, 158
Biosafety in Microbiological and Biomedical Laboratories (BMBL), 46
biosafety levels or BSL, 43, 70, 152, 153, 155, 159, 167
biosecurity, xiv, 44, 46, 77, 80, 93, 118
biosurveillance, 63, 66, 72, 77, 78, 82–83, 86, 93, 118, 166
biotechnology : advanced technologies, 30; advanced manufacturing, 33, 41; DNA engineering, xiii, 30, 38, 41, 88, 97, 160; definition of, 1, 8, 32, 33; economic importance of, xiv, xvi, 30, 32, 36, 45, 48, 49, 64–67, 93, 97, 99, 151, 164; information technology, 30, 33, 35, 44, 50, 64, 97; nanotechnology, 33, 41, 97, 174; neuropharmacology, 33, 41; problem, 30–35, 161–163; rate of change, xiii, xiv, xv, 29–52, 161–163; strategic crossroads with national security and arms control, xv; trends, 64
bioterror, xi, xii, 16, 24, 36, 38, 41, 44, 49, 50, 51, 54, 55, 58, 59, 60, 62, 63, 69, 70–71, 73, 77, 79, 82, 86, 99, 115, 116, 117, 121, 126, 127, 145, 147, 148, 149, 155, 161, 165, 175, 195
Bioweapons Prevention Project (BWPP), 122, 125, 171
Bolton, John, 116, 117, 119
bombers, 3
Brazil, 59
Brezhnev, Leonoid, 3
bubonic plague, 47, 49, 56, 77, 137, 140, 141
Bush Administration, 157
Bush, George W., 137, 157
Bush, Vannevar, 57
Butler, Mark, 69
Biological Warfare (BW) testing, 13, 16, 17, 18, 26, 74; Japanese, 23; Large Area Coverage or LAC, 17; New York, 17,

55; Project 112, 17, 18, 20; San Francisco, 17; Soviets, 21–22, 56, 74, 136; Sites for U.S. testing, 13, 16, 137; United Kingdom, 77; U.S. conclusions, 17–20; Washington, DC, 17; zinc cadmium sulfide, 17

BW, importance of knowledge, 8, 18, 23, 36, 43, 91, 92, 147

BW, poor man's atomic weapon, xiii, 55

BWC, Articles of the Convention,: Article I, 6, 8, 15, 48, 87–90, 91, 92, 109, 112, 126, 139, 142, 161, 164, 173, 180; Article II, 90–91, 139, 180; Article III, 91–92, 164, 180; Article IV, 92–94, 181; Article V, 94–95, 106, 109, 110, 111, 181; Article VI, 94–95, 102, 106, 181; Article VII, 95–96, 181; Article VIII, 99–100, 181; Article IX, 100, 182; Article X, 30, 91, 92, 96, 98, 98–99, 107, 110, 111, 118, 120, 122, 163, 164, 173, 175, 182; Article XI, 100, 182; Article XII, 100, 103, 182; Article XIII, 101, 183; Article XIV, 101, 183; Article XV, 101, 184

BWC, entry into force, 1, 12, 27, 30, 41, 47, 48, 49, 52, 59, 68, 90, 94, 102, 105, 145, 149, 166, 180, 182, 183

California, 17, 105

Cambodia, 73, 107

Canada, 14, 17, 71

cancer, 35, 37, 41, 45, 79

Carlson, Robert, 64, 70, 97

Centers for Disease Control and Prevention (CDC), 54, 152, 158

Central Intelligence Agency (CIA), 12, 58, 90, 94

Chairman of the Joint Chiefs of Staff (CJCS), 12

chemical agents,: ammonia, 147; chlorine, 147; nerve agent, 13, 147; sheep incident, Dugway Proving Ground, 13

Chemical Plants : at Berdsk, 138; at Omutninsk, 138

Chemical Weapons Convention (CWC), 4, 27, 76, 108, 129, 197; Article I, 4–5; comparison to BWC, 6, 7, 27, 85, 100, 133, 152, 161, 163, 169, 198; convergence of BWC and CWC, 7, 33,

42, 85, 88; destruction, 76, 163; entry into force, 100; members, 100, 102, 154; Organization for the Prohibition of Chemical Weapons (OPCW), 5, 132, 169, 176; review conferences, 5; schedules, 76, 168; verification and inspections, 4, 7, 42, 100, 131, 133, 159, 163, 167, 168, 198, 199

chemical weapons destruction facilities (CWDFs), 167

chemical weapons production facilities (CWPFs), 167

chemical weapons storage facilities (CWSFs), 167

Chemical, Biological, Radiological, Nuclear (CBRN), 2, 147, 149, 191

Chesapeake watershed, 38

China, 32, 58, 59, 65, 94, 96, 105, 120, 168, 173

cholera, 9, 23

circular error probable (CEP), 75

classic swine fever, 65

Clingendaal, 120

Clinton Administration, 89

Clinton, Hilary, 122, 133, 166, 176

Clostridium perfringens , 141

Cold War, xi, 2, 5, 56, 75, 110, 130, 146, 190, 191

Colorado, 27

Commission on the Prevention of Weapons of Mass Destruction Proliferation and Terrorism (commonly known as the Graham/Talent WMD Commission), 73

committee of the whole, 104

Conference on Disarmament (CD), 14, 76, 86, 169

compliance, xiii, 1, 2, 128, 129, 155, 171; BWC, xiii, xv, 6, 57, 58, 81, 82, 84, 86, 87, 93, 94–95, 102, 106, 107, 108, 109, 110, 111, 114, 115, 123, 125–144, 148, 150, 155, 156, 159, 160, 161, 171, 176; CWC, 4, 167, 176; Iraq noncompliance with BWC, 140–142, 168; measures, 3, 131–136, 143, 155, 160, 161, 168, 190, 193–194; nuclear treaties, 82; Soviet noncompliance with BWC, 136–139; verification, compliance contrast with, 129, 155, 171, 190

Compliance Review Group (CRG), 158

Comprehensive Test Ban Treaty (CTBT),
4
computer-aided design (CAD), 33
confidence building measures (CBMs),
xvi, 47, 48, 62, 63, 85, 89, 91, 102, 109,
111, 114, 118, 126, 127, 131, 133, 143,
154, 160, 161, 162, 166, 167, 168, 169,
173, 187–188
Congress, U.S., 3, 7, 10, 12, 13, 15, 25, 27,
54, 58, 94, 98, 129, 130, 158, 201
Congressional Office of Technology
Assessment, 54
Congressional Research Service (CRS),
129
consequence management, 74, 75
consequence of bioevent, 81, 100, 123,
147–148, 174; accident of biotech
misuse, 37, 69, 70; economic, 65, 66,
67, 78, 148; public health, 68;
scenarios, 73; societal, 67, 78, 148;
unpredictable BW weapons, 26;
zoonotic, 49, 69
Convention on Cluster Munitions, 4
Conventional Armed Forces in Europe
(CFE) Treaty, 4, 128, 133
Cooperative Threat Reduction (CTR)
program, 2, 22, 56, 67, 90, 164, 191
Counterproliferation, 74, 75, 195
cowpox (see also *vaccinia*), 9
Crick, Francis, 30
Cuban Missile Crisis, 2, 18, 189
cyber, 42, 43

de novo synthesis, xiii, 30, 38, 160
Defense Threat Reduction Agency
(DTRA), 55, 70
democratization of biology, 36
Denmark, 71
Department of Agriculture, 27, 138
Department of Defense, 13, 20, 55, 68, 77,
92, 164, 166
Department of Health and Human Services
(HHS), 80
Department of Health, Education, and
Welfare, 27
Department of Homeland Security, 72,
138, 158, 166
Department of State, 139, 142
diagnostics, 39

disease: accidental release potential, 69,
98; BWC role, 82, 83, 84, 96, 110, 117,
118, 121, 147, 155, 165, 175, 182, 187,
202, 204; concept of, xi, 7, 8–9, 10, 18,
31–32, 53, 68, 128, 147, 174; emerging
infectious disease, 63, 65, 66, 67, 69;
fear of, 24, 58, 149; fighting disease,
61, 62–63, 77, 78, 82, 83, 118, 166;
foreign animal disease, 51, 57, 62, 63,
65; globalization effect on, 49, 69;
immunology, 9, 18, 39–40, 41, 77, 79,
128, 174; impact of, 65–66, 69; Koch's
principles, 9; naturally occurring versus
BW, xiv, 7, 16, 18, 20, 22, 24, 40, 47,
53, 55, 68, 84, 149, 174; overwhelming
dose, 7, 55
DIY biology, 36, 41, 43, 44, 70, 97
DNA (deoxyribonucleic acid), 30, 31–32,
33
Dual Use Research of Concern (DURC),
98, 158
dual-use, xiv, 41, 49, 50, 51, 61, 75, 78,
79–81, 89, 91, 92, 93, 98, 125, 127,
142, 144, 152, 153, 158, 159, 174, 176,
185
DuBridge, Lee, 15
Dugway Proving Ground, 13, 137

East Germany, 15
Ebola virus, 7, 49, 63, 152
Egypt, 58, 66, 78, 94
Eisenhower Administration, 10
Eisenhower Administration, Basic
National Security Strategies or BNSPs,
10
Eisenhower, Dwight D., 10, 17
Ekeus, Rolf, 141
Emergency Use Authorization (EAU), 73
encapsulization, 41
Encyclopedia of DNA Elements
(ENCODE), 31, 32
entry into force (see BWC, entry into
force)
environmental concerns, xiv, 35, 41, 53,
75, 76, 77
Erasmus Medical Center, 80
Escherichia coli , 35, 37, 71, 165
Europe, 4, 9, 23, 34, 56, 86, 121, 128, 133,
165

Evans Medical Limited in Liverpool, 138
executive orders (EO), 156
experts meeting, xiv, 161, 163

Federation of American Scientists (FAS), 10, 41, 97
Ferment project, 136
Fink Report, 46, 79
fissile material, acquisition, 81, 132, 147, 159, 162, 175, 176
flexible response, 11
Florida, 38
Food and Agriculture Organization of the United Nations (FAO), 62, 63, 96, 110, 121, 170, 175; Article X, 118; relationship with BWC, 61, 62, 63, 77, 146, 159, 162
food security, xiv, 62, 75, 78, 151, 174, 185
Foot and Mouth Disease (FMD), 51, 65, 66; Korea outbreak, 65; United Kingdom outbreak, 65
Ford, Gerald, 3, 27
former Soviet Union, 4, 89, 90, 164
Fort Detrick, Maryland (also Camp Detrick), 10, 16, 27, 55, 137, 166
Fouchier, Ron, 80
France, 23, 105, 149
Francisella tularensis , 16, 19, 55
Frankfurt Airport, Germany, 165
Fulbright, J. William, 12
fungi, 7

G-8 Global Partnership, 164
genetically modified organisms (GMO), xii, 34–35, 41, 49, 88, 109
Geneva Protocol, 4, 13, 14, 26, 27, 179
Geneva, Switzerland, 6, 86, 99, 104, 105, 106, 108, 109, 110, 119, 165, 166, 169, 179, 181, 182, 195
Georgia, 164
German high-explosive buzz bombs, 16
Germany, 15, 61, 71, 165
glanders bacteria, 56
globalization, 49, 53, 63, 69, 76, 174
Gore, Albert, 138
Government Accountability Office (GAO), 152
Grand Bazaar, Istanbul, 165

Gromyko, Andrei, 14, 15
gross national product (GNP), 1, 64, 65, 98
Groton, Connecticut, 138
Gruinard Island, 77
Gulf War, 59, 74

H5N1 avian influenza, 43, 44, 48, 70, 80, 81, 98, 127, 148, 158, 165
Hague, Netherlands, 5, 162
Hendra virus, 63
heroin with anthrax contamination, 72
HIV/AIDS, 51, 67, 109
Homeland Security Presidential Decisions (HSPD), 156
Hooke, Robert, 30
host susceptibility, 8
Human Genome Project, 35, 41
Hungary, 23, 111
Hussein, Saddam, 24, 59, 74, 110, 140
Huxsoll, David, 98, 141

IL-4, also Interleuken-4, 37, 128
immune system, 9, 18, 39–40, 41, 77, 79, 128, 174
Implementation Support Unit (ISU), 71, 72, 86, 118, 121, 122, 123, 143, 154, 162, 163, 165–170, 173, 175
India, 32, 59, 60, 65, 120, 168, 173
Indonesia, 32
Industrial Age, xiv, 41, 42
influenza, 9, 32, 38, 39, 44, 45, 47, 48, 49, 62, 66, 70, 77, 78, 80, 81, 82, 98, 127, 148, 152
Information Age, 42, 64
Inspire magazine, 60
Institute of Molecular Biology in Koltsova, 137
Institute of Ultrapure Preparations in Leningrad, 137
intellectual property (IP), 30, 45, 67, 97
Intelligence Community (IC), 13
Interagency, 12, 13, 26, 120
Intermediate Range Nuclear Force (INF) treaty, 3, 130
Internal Health Regulations (IHR), 63, 71, 82, 93
International Atomic Energy Agency (IAEA), 4, 82, 131, 132, 159, 162, 163, 175, 197

International Plant Protection Convention (IPPC), 96, 118
Internationally Genetically Engineered Machine (iGEM), 36, 41
INTERPOL, 72, 83, 162, 165
intersessional meetings, 52, 95, 105, 118, 122, 161
invasive species, 11, 38, 62, 78
investigational new drug (IND), 73
Iran, 24, 55, 58, 59, 68, 94, 116, 120, 140, 173
Iraq, 24, 59, 74, 90, 95, 110, 115, 116, 140–142, 144, 151, 159, 168
Israel, 58, 94
Ivins, Bruce, 69

J. Craig Venter Institute, 30, 38, 40
Janssen, Zacharias, 30
Japan, xii, 15, 23, 59, 65, 72
Jenner, Edward, 9
Johnston Atoll, 19
Joint Chiefs of Staff (JCS), 12, 13, 19, 20, 26
Joint Compliance and Inspection Commission (JCIC), for START, 132
junk DNA, 31–32

Kamal, Hussein, 141
Kastenmeier, Robert W., 10
Kawaoka, Yoshihiro, 80
Kazakhstan, 21, 90, 164
Kennedy Administration, 11
Kennedy, John F., 18
Killian, James, 17
killing winds, 55
Ki-moon, Ban, 104
Kissinger, Henry, 12, 13, 14, 15, 190
Kistiakowsky, George, 17
Koch, Robert, 9
Koch's principles, 9
Korean War, 10, 16
knowledge, importance of for BW development, 8, 18, 23, 36, 43, 91, 92, 147

Laird, Melvin, 12, 13, 26
Laos, 73, 107
Large Area Coverage or LAC, 17
Latin America, 4, 146

Law of the Sea Treaty, 76, 177
lethal dose 50% or LD50, 7, 40
Lederberg, Joshua, 11
Lego, 37
Lemon-Relman Committee, 38
light bulb (as weapon), 17, 55
London, 7, 82, 149, 184
Los Angeles, 165

Mahley, Donald, 115, 116, 117, 195–205
malaria, 51, 67
Malaysia, 32
Manila, 120
Marburg virus, 47, 63
Markov, Georgi, 149
Maryland, 16, 27
McCarthy, Joseph, 12
medical countermeasures, 8, 16, 26, 54, 73, 153, 156
medicine, xiii, 1, 9, 32, 35, 39, 41, 44, 61, 64, 76, 79, 92, 98, 151, 152
meeting of state parties, 6, 86, 95, 119, 154, 157
Merck Report, 9
Merck, George W., 9, 16
Meselson, Matthew, 13
metabolomics, 33, 41, 47
Mississippi, 16
monoclonal antibodies, 41
Montreux, 120
Moore's Law, 64
Moscow, 7, 184
mousepox, 37, 69
multiple integrated reentry vehicles (MIRV), 3, 75
mycotoxin, 107, 136

Namibia, 59
nanotechnology, 33, 41, 97, 174
National Academy of Science (NAS), 59
national implementation, xiii, xvi, 6, 82, 86, 91–94, 109, 114, 118, 120, 133, 147, 148, 155, 158, 159, 173, 207–209
National Institutes of Health (NIH), 69
National Science Advisory Board for Biosecurity (NSABB), 46, 80
National Security Advisor, 14, 15
National Security Council (NSC), 10, 14, 56

National Security Decision 35, 27
National Security Decision 44, 27
National Security Study Memorandum
  (NSSM) 59, 12–15, 19, 20, 23, 25, 26
National Security Study Memorandum
  (NSSM) 85, 26
*National Strategy for Countering
  Biological Threats*, see also Presidential
  Policy Directive (PPD)-2, 54, 119, 156,
  157
national technical means (NTM), 3, 130,
  131, 133, 193
nerve agent, 13, 147
Netherlands, 5, 71, 80, 122
neuropharmacology, 33, 41
neutron bomb, 55
New START treaty, 132, 133, 151, 177
New York, 17, 30, 55, 138, 165
Newcastle disease, 65
Nipah virus, 63
Nitze, Paul, xvii, 130, 135, 155, 161
Nixon Administration, 13
Nixon, Richard M., xii, xv, 9, 12, 26, 27,
  46
Nobel prize, 11, 36, 50
non-aligned nations (NAM), 67, 91, 145
nonproliferation, 2, 12, 57, 60, 61, 74, 75,
  191
North Atlantic Treaty Organization
  (NATO), 4, 10, 146
North Korea, xii, 58, 59, 68, 94, 116
Nuclear Nonproliferation Treaty (NPT), 4,
  12, 27, 75, 102, 131, 132, 151, 154,
  159, 161, 197
Nuclear Weapon Free Zones (NWFZ), 4
Nunn-Lugar Cooperative Threat Reduction
  Act, 22, 23, 56, 67, 90, 164

Obama Administration, 56, 119, 120, 156,
  157
Obolensk, 56, 137, 138
Office of Technology Assessment, 25, 54,
  94
One Health, 60, 62, 72, 96
on-site inspections, 2, 106, 131, 132, 133,
  135, 159, 170, 193, 197
Operation Iraqi Freedom, 75
Operation White Coat, 16
Oregon, xii, 59

Organization for the Prohibition of
  Chemical Weapons (OPCW), 5, 132,
  169, 176
Organization of the United Nations (FAO),
  61, 62, 63, 77, 96, 110, 118, 121, 146,
  159, 162, 170, 175
Outer Space Treaty, 4

Pasechnik, Vladimir Artemovich, 21, 56,
  137, 191
Pasteur, Louis, 9
Patrick, William, 50, 55
Penn Station, New York, 165
Perestroika, 22
perimeter portal continuous monitoring
  (PPCM), 131, 132, 133
Pfizer, 138
Pine Bluff, Arkansas, 16, 137
Ping Fan facility, 23
plague, 23, 47, 49, 56, 77, 105, 123, 137,
  140, 141
Plum Island Animal Disease Center, 138
Poland, 71
poliovirus, 30, 152
polymerase chain reaction (PCR)
  amplification, 36, 47, 48, 51
poor man's atomic weapon, xiii, 55
Popov, Sergei, 56
potato beetle, 65
preparatory conference (PrepCon), 104
Presidential Policy Directives (PPD), 156;
  Presidential Policy Directive (PPD) – 2,
  56, 119, 156; Presidential Policy
  Directive (PPD) – 8, 54
President's Science Advisory Committee
  (PSAC), 14, 17
PRIICs (Pakistan, Russia, India, Iran and
  China), 120–121, 173
Project 112, 18, 20; Night Train, 18; Red
  Cloud, 18; Shady Grove, 18, 19;
  Speckled Start, 18; West Side, 18
Project Bacchus, 55, 56, 70, 126
Project Clear Vision, 89
Proliferation Security Initiative (PSI), 61,
  75
proteomics, 33, 40, 41, 47
Protocol for the Prohibition of the Use in
  War of Asphyxiating, Poisonous or
  Other Gases and of Bacteriological

Methods of Warfare, 4, 9, 12, 85, 99, 103, 146, 179, 181

public health, xiv, xvi, 1, 61, 63, 64, 68, 72, 75, 76, 77, 81, 82, 92, 93, 94, 95, 96, 98, 122, 151, 156, 166, 175, 185

Pugwash Conference (also known as the Conference on Science and World Affairs), 10

Putin, Vladimir, xii

Q fever, 47

Rafsanjani (Hojjat al-Islam Akbar Hashemi-Rafsanjani), 55

Rajneeshee cultists, xii, 59, 152

recombinant DNA, 30, 46, 47, 105

Review Conferences, xiv, xvi, 29, 30, 76, 85, 103–105, 123–124; First, 1980, 106–107; Second, 1986, 107–110; Third, 1991, 110–111; Fourth, 1996, 114–115; Fifth, 2001 & 2002, 115–117; Sixth, 2006, 118–119; Seventh, 2011, 119–122

Review Conference, BWC and CWC compared, 6

ribonucleic acid (RNA), 31, 32, 39

rice blast, 47, 65

ricin, 7, 149

rickettsia, 7

Rift Valley Fever (RFV), 63

rinderpest, 65

RNA interference, 39, 41

Roosevelt, Franklin D., 9, 16, 20, 26

Rosebury, Theodor, 10

Rotterdam, 165

Russia,: 2011 review conference participation, 120; arms control, 2, 177, 191; BWC compliance, 139; BW program, 58, 59, 94, 136, 139; CWC destruction, 5; former Soviet Union, 90; Sverdlovsk admission, 108; verification position, 150, 173; See also Soviet Union

Salk Institute, 137

San Francisco, 17

saxitoxin, 7, 88, 90

schedules, CWC, 4, 76, 168

Schmallenberg virus, 62, 153

Secretary of Defense, 12, 19, 20

select agent, 61, 126, 152, 159

Senate Foreign Relations Committee (SFRC), 130

Senate, U.S., 3, 7, 25, 27, 58, 94, 98, 201

sensors, 41, 82, 97, 135

sequencing, 31–32, 33, 35, 37, 38, 39, 51, 64, 97, 128, 152

Serratia marcescens, 17, 55

Seventh Day Adventists, 16

Severe Acute Respiratory Syndrome (SARS), 38, 44, 49, 66, 69, 96

Shigella dysenteriae, 148

Shiro, Ishii, 23

simulants, use of, 17

Singapore, 32, 65

smallpox, 8, 9, 47, 49, 56, 69, 73, 78, 137, 152, 165

Smith, Gerard, 26

Snow, John, 9

South Africa, 58, 59, 94

Southeast Asia, 11, 107, 136, 164

Soviet Union, 58, 94, 146; arms control, 2, 3, 4, 75, 130, 189; biotechnology, 109; BWC leadership, 86, 149, 183; BW offensive program, xii, 15, 20–23, 47, 56–57, 58, 89, 90, 105, 136, 146, 149; BW sites, 22, 90, 106; experimentation with BW, 22, 89; mistrust between U.S. and Soviets, 2, 14, 106, 107–108, 130, 137–139; noncompliance, xii, 57, 59, 105, 108, 110, 136–137, 159; testing, 21–22, 56, 74, 136; verification, 73, 88, 108

special verification commission (SVC), for INF treaty, 132

spectrum of biological threats, xiv, 67, 68–71, 73, 147, 148, 157, 185

St. Louis University, 69

stabilization, 18, 41, 79, 92, 97

Standing Consultative Commission (SCC), for ABM Treaty, 132

Stanford University, 11

Staphylococcus enterotoxin B (SEB), 16, 19, 47, 55

Star Wars, 104

START I, 3, 131, 133

State Research Center of Virology and Biotechnology (known as "Vector"), 56

State University of New York, 30
Stepnogorsk, 21
Stern, Jessica, 58
Strategic Arms Limitation Talks (SALT), 3, 130
Strategic Arms Reduction Talks (START), 3, 131, 132, 133
Strategic Offensive Reductions Treaty (SORT), 3
Sverdlovsk, 22, 57, 73, 106, 107, 108, 136, 137, 149, 191
Sweden, 14, 71, 106
synthetic biology, 7, 25, 33, 34, 36, 37–38, 47, 64
Syria, xi, 58, 59, 68, 94

Tauscher, Ellen, 119
Terre Haute, Indiana, 16
The Dead Hand, 57
The Marshall Plan, 18
Toth, Tibor, 111, 113, 114, 117
treaty limited equipment (TLE), 129, 131
Treaty of Pelindaba, 4
Treaty of Tlatelolco, 4
Tripartite Agreement, 137–139
tuberculosis, 51, 67, 148
tularemia, 8, 16, 19, 47, 56, 140
typhoid, 23
typhus, 47

U.S. Armed Forces, 10, 13, 20, 133, 138
U.S. Army Chemical Corps, 11
U.S. decision to renounce BW, xii, 9, 12–15, 19–20, 25–27
Ukraine, 90, 164
Union of Soviet Socialist Republics (USSR), see Soviet Union
Unit 731, 15, 23
United Kingdom, 14, 21, 23, 59, 65, 72, 105, 106, 137, 138–139, 149, 183
United Nations, xvi, 14, 15, 61, 73, 74, 76, 105, 108, 177; Charter, 83, 94, 96; United Nations Disarmament Commission (UNDC), 135; United Nations General Assembly (UNGA), 15; United Nations Mission in the Palais des Nations, 104, 165; United Nations Secretary General (UNSYG), 106; United Nations Security Council,

72, 94, 101, 105, 108; United Nations Security Council Resolution (UNSCR) 687, 140; United Nations Security Council Resolution (UNSCR) 1540, 74, 95, 118; United Nations Special Commission (UNSCOM), 95, 140–142, 144, 151, 168; United Nations Monitoring, Verification and Inspection Commission (UNMOVIC), 95, 168
United States Army Medical Research Institute for Infectious Diseases (USAMRIID), 69, 98, 137, 141, 166
University of Wisconsin, 81
Utah, 13, 16
Uzbekistan, 90, 164

vaccines, 16, 26, 27, 32, 41, 45, 67, 68, 73, 79, 80, 87, 88, 97, 98, 127, 140, 152, 188
van Leeuwenhoek, Antony, 30
Variola major (see smallpox)
Venezuelan equine encephalitis (VEE), 19, 47, 56
Venter, Craig, 88, 152
Verification Experts (VEREX), 111, 112–113
verification protocol for BWC, xiii, xv, 84, 94, 163, 167; avoiding discussion at 2011 review conference, 120; comparison to CWC, 6, 163; concerns with, 81, 136, 139, 151; confidence building measures relationship, 160, 161, 163; developing a protocol, xvi, 86, 120, 131, 132, 139, 143, 146, 150, 154, 155, 160, 167, 168; impact of not having, 106, 108, 120, 121, 159, 161, 162; U.S. position, 115, 121, 134, 139, 142, 150, 166, 190; U.S. transparency visit, 176
verification,: arms control, 2, 102, 125, 129–130, 134, 135; BWC lack of verification mechanisms, xvi, 6, 7, 10, 15, 57, 58, 84, 86, 88, 101, 106, 107, 150, 152; CWC, 4–5, 42, 76, 100, 133, 167–168; compliance versus verification, 123, 129, 134, 135, 155–156, 160, 176, 190; costs of verification, 135, 167, 170; debate related to BWC, xvi, 71, 88, 102, 105,

106, 120, 121, 124, 125, 128, 134, 146, 147, 166, 167, 171, 176; definition, 129, 130, 134, 168; difficulty in assessing noncompliance, 142; difficulty in verifying BW, 111, 136–142; effective, xvii, 84, 130, 130, 161, 168, 176; international perspectives on BWC verification, 109, 110, 111, 118, 119, 176; limits of, 143–144; means, 3, 94, 102, 106, 109, 112–113, 128, 131–132, 150, 161; nuclear arms control, 3, 133, 159; perfect, 135, 168, 176; principles, xvi, 135, 193–194; U.S. concerns about BWC verification, xi, xvii, 88, 94, 108, 120, 121, 122, 123, 150, 151, 159

Verification, Research, Training and Information Centre (VERTIC), 82, 158, 207–209

Vibrio cholerae, 9

Vietnam War, 11, 12, 26, 85, 107

Vigo, Indiana, 138

virulence, 31, 40, 48, 80, 81, 97, 127

Vladivostok, 3

Vozrozhdeniye (Rebirth) Island, 22

War Reserve Service (WRS), 16

Warsaw, 165

Warsaw Pact, 4, 10, 15, 108, 146, 184

Washington, D.C., 7, 17, 55, 56, 66, 196

Washington Post, 89, 126

Watson, James, 30

weapons of mass destruction (WMD), 2, 21, 57, 61, 81, 85; Commission on the Prevention of Weapons of Mass Destruction Proliferation and Terrorism (commonly known as the Graham/ Talent WMD Commission), 73; Iraq, 140; *Militarily Critical Technologies List Part II: Weapons of Mass Destruction Technologies* "Biological Weapons Technology", 68; official documents concerning, 156

Webster, William, 58, 94

West Nile virus, 49

wheat rust, 65

Wheeler, Earle, 20, 26

white powder hoaxes, 59, 83

Wilton Park, 120

Wimmer, Eckard, 152

World Health Organization (WHO), 61, 62, 63, 71, 72, 77, 82, 83, 96, 166; Article X, 110, 118; BW scenario, 54; International Health Regulations (IHR), 63, 71, 93; membership, 82; relationship with BWC, 61, 63, 121, 146, 159, 162, 165, 170, 175, 187

World Organization for Animal Health (OIE), 61, 62, 63; Article X, 118; at 2011 review conference, 96; relationship with BWC, 61, 62, 63, 77, 110, 121, 146, 159, 162, 170, 175

World War I, 15, 23, 24

World War II, 2, 9, 10, 15, 16, 74, 189

Xenotropic Murine Leukemia Virus-Related Virus (XMRV), 37

Yellow Rain, 73, 107, 108, 136, 137, 190

Yeltsin, Boris, 22, 136

Zhukov, Georgiy, 57

About the Author

Dr. Daniel M. Gerstein is a security and defense professional who has served in a variety of positions as a Senior Executive Service (SES) government civilian, in uniform, and in industry. He is currently serving as the Deputy Under Secretary for Science & Technology in the Department of Homeland Security. He is also an adjunct professor at American University in Washington, D.C., at the School of International Service (SIS), where he teaches graduate-level courses on biological warfare and the evolution of military thought.

Before joining DHS, he served as Principal Director for Countering Weapons of Mass Destruction (WMD) within the Office of the Secretary of Defense (Policy). In uniform, he has served on four different continents participating in homeland security and counterterrorism, peacekeeping, humanitarian assistance, and combat, in addition to serving for over a decade in the Pentagon in various high-level staff assignments. Following retirement from active duty, Dr. Gerstein joined L-3 Communications as Vice President for Homeland Security Services, leading an organization providing WMD preparedness and response, critical infrastructure security, emergency response capacity, and exercise support to U.S. and international customers.

He has been awarded numerous military and civilian awards, including an award from the government of Colombia, the Department of State's Distinguished Service Award, and the U.S. Army Soldier's Medal for heroism. He has published numerous books and articles on national security, biological warfare, and information technology, including *Bioterror in the 21st Century* (Naval Institute Press, October 2009); "ICMA Report: Planning for a Pandemic," *ICMA Press* 39, no. 3 (2007); *Securing America's Future: National Strategy in the Information Age* (Praeger Security International, September 2005); *Leading at the Speed of Light* (Potomac Books, November 2006);

Assignment: Pentagon: How to Excel in a Bureaucracy (Potomac Books, May 2007). He has also served as a fellow at the Council on Foreign Relations and is a current member.

Dr. Gerstein holds degrees from the United States Military Academy, Georgia Tech, National Defense University, U.S. Army Command and General Staff College, and George Mason University. He resides in Alexandria, Virginia, with his wife, Kathy. They have two daughters.